Lectures on

SUPERMANIFOLDS, GEOMETRICAL METHODS & CONFORMAL GROUPS

Published by

World Scientific Publishing Co. Pte. Ltd.,
P O Box 128, Farrer Road, Singapore 9128
USA office: 687 Hartwell Street, Teaneck, NJ 07666
UK office: 73 Lynton Mead, Totteridge, London N20 8DH

LECTURES ON SUPERMANIFOLDS, GEOMETRICAL METHODS AND CONFORMAL GROUPS

ISBN 9971-50-808-7

Printed in Singapore by JBW Printers & Binders Pte. Ltd.

Lectures on

SUPERMANIFOLDS, GEOMETRICAL METHODS & CONFORMAL GROUPS

given at Varna, Bulgaria

Edited by: **H.D. Doebner, J.D. Hennig** ***(Clausthal)***
T.D. Palev ***(Sofia)***

World Scientific
Singapore • New Jersey • London • Hong Kong

Preface

This lecture notes volume contains selected and updated material presented at the International Workshop Program on Quantum Field Theory and related mathematical topics organized in Varna, Bulgaria, jointly by the Institute for Nuclear Research and Nuclear Energy (INRNE), Bulgarian Academy of Sciences, Sofia, and Arnold Sommerfeld Institute for Mathematical Physics (ASI), Technical University of Clausthal, FRG.

The selection is geometrically biased and is centered around the following topics:

I. Supermanifolds

II. Differential Geometric Methods

III. Conformal Groups.

Contained among others are Manin's and Penkov's paper on the formalism of Left and Right Connections on Supermanifolds, Uhlmann's algebraic view on the Differential Geometry of Smooth Manifolds and an elaborate contribution by I.T. Todorov et al. on the Conformal Group, its representations and its application to Quantum Field Theory, Conformal Composite Fields and Operator Product Expansions. The articles of Uhlmann and those of Todorov have partly a review character and should be useful for readers, who want to enter the field.

Geometrical methods for a mathematical modelling of complex physical systems are now well established and successful, especially in Quantum Field Theory. The field is in general well and quickly covered in journals and proceedings. These lecture notes on specific topics try to bridge the gap to detailed chapters of advanced textbooks in connection with short research notes as a kind of example.

Acknowledgements

We wish to express our gratitude to the following organizations for generous finanical support and for other assistance:

— Institute for Nuclear Research and Nuclear Energy (INRNE), Bulgarian Academy of Sciences, Sofia (Bulgaria).

— Technical University of Clausthal (FRG), Arnold Sommerfeld Institute for Mathematical Physics.

We are thankful to Academician Prof. Christov for his encouragement and his support. We want to thank World Scientific, Singapore, for assisting us. We acknowledge the work of the secretarial staff, especially of Barbara Buck and of J. Lopez-Fenner (Clausthal) for retyping some of the manuscripts.

H. D. Doebner
J. D. Hennig
T. D. Palev

TABLE OF CONTENTS

I. Supermanifolds

The Formalism of Left and Right Connections on Supermanifolds

Yu. I. Manin, I. B. Penkov
Steklov Mathematical Institute, USSR Academy of Sciences
117956 Moscow GSP1, USSR

Introduction

This paper originated in the authors' attempts to understand characteristic classes in supergeometry.

We start with recalling one of the constructions of characteristic classes in the category of C^∞- manifolds, [BC], [MS], [W]. Let X be a C^∞- manifold, $\mathcal{F}$ a locally free sheaf (the sheaf of sections of a vector bundle) on it. The characteristic classes of $\mathcal{F}$ in the de Rham cohomology can be constructed as follows. Choose a connection ∇ on $\mathcal{F}$ and consider its curvature $F_\nabla \in \Gamma(X, End\mathcal{F} \otimes \Omega^2 X)$. The forms $trF^i_\nabla \in \Gamma(X, \Omega^{2i}X)$ are closed and their cohomology classes do not depend on the choice of ∇. If the structure group of $\mathcal{F}$ is reduced, instead of the traces of F^i_∇ one should consider all invariant polynomials on the Lie algebra of the structure group. In the complex-analytic situation a global connection may not exist, instead one should first construct the Atiyah-class F^{an}_∇, which is the obstruction to the existence of a holomorphic connection on $\mathcal{F}$,$F^{an}_\nabla \in H^1(X, End\mathcal{F} \otimes \Omega^1 X)$, and then consider $tr[(F^{an}_\nabla)^i] \in H^i(X, \Omega^i X)$.

In supergeometry one may carry out these constructions too. However, we were not satisfied with both of them because the cohomology classes one obtains in this way have "pure even dimension", being possible to integrate differential forms only along chains of odd dimension zero. But in [BL1] Bernstein and Leites introduced objects, which can be integrated along chains of odd codimension zero. These are the integral forms. It was natural to conjecture that the complex of integral forms is connected with some "integral sequence of a sheaf", in the same manner as the de Rham complex leads to the de Rham sequence of a sheaf with connection.

The aim of our paper is to present this new construction. We introduce the notion of a right connection on a vector bundle and discuss its integral sequence (we call it the Spencer sequence of the right connection).

Although the curvature of a right connection is again a differential form, we are able to produce characteristic classes with values in integral forms by a trick explained in §4.

A way out of the purely even construction is suggested by another remark of Bernstein and Leites. One may integrate the so called pseudodifferential forms, fastly decreasing in the differentials of the odd coordinates, [BL2]. In our case they can be constructed as power series of the form $\sum_0^\infty a_i trF^i_\nabla$.

The problem of characteristic classes in supergeometry is closely connected with the problem of the (co)homology theory, which carries them. The approach via classifying spaces gives a possible definition of "Schubert supercells" with varying odd dimension.

For example in the projective space $\mathbf{P}^{n|m}$ lie $\mathbf{P}^{a|b}$ with all possible a and b: $0 \le a \le n$, $0 \le b \le m$, which may be given with $n-a$ even and $m-b$ odd equations. However it is not clear in what homology theory there are classes of these cells, with fewer relations between them, than in the case $m=0$.

An idea of A. S. Schwarz to realize characteristic classes not by forms, but by densities, seems quite interesting too. Densities are objects which can be integrated along chains of arbitrary codimension, but they do not have a natural structure of a complex, and have some unusual properties.

§1. Preliminaries

1.1. Let $X = (\mathcal{K}, O_x)$ be a supermanifold over k, where by k we denote one of the fields $\mathbb{R}$ or $\mathbb{C}$. This means that $\mathcal{K}$ is a topological space and $O_x = O_0 \oplus O_1$ - a sheaf of supercommutative rings with the following properties:

1. If we denote by $N \hookrightarrow O_x$ the ideal of nilpotents, then the ringed space $X_{red} \overset{\text{def}}{=} (\mathcal{K}, O_x/N)$ is a C^∞- real or analytic (real or complex) manifold.

2. The O_x/N -module N/N^2 is locally free, $rkN/N^2 < \infty$ and O_x is locally isomorphic to the Grassman algebra $\Lambda^{\cdot}(N/N^2)$. In this paper we consider parallelly C^∞- or analytic supermanifolds over k. When it is necessary, we restrict ourselves with the smooth or analytic category.

Given a supermanifold X, there is always an embedding $g_X : X_{red} \hookrightarrow X$, but in the analytic category there may be no inverse projection. However the following is true:

Theorem (Batchelor, [B]): In the C^∞ -case there always exists an isomorphism $X \simeq (\mathcal{K}, \Lambda^{\cdot}(N/N^2))$.

If $\mathcal{F}$ is a locally free O_x -module of rank $p \mid q$, the modules $\prod \mathcal{F}$ and $\mathcal{F}^*$, of ranks $q \mid p$ and $p \mid q$, are defined. $\prod \mathcal{F}$ is "$\mathcal{F}$ with the opposite parity", and $\mathcal{F}^*$ is the dual of $\mathcal{F}$, [L].

By TX we denote the tangent sheaf of X, which is locally free of rank $dimX$. In this paper we set $dimX = n \mid m$ where $n \overset{\text{def}}{=} dimX_{red}$ and m is by definition equal to $rk_{O/N}N/N^2$.

1.2. The algebra of differential forms on X is $\Omega^{\cdot}X \overset{\text{def}}{=} \mathbf{S}(\prod TX^*)$, where by $\mathbf{S}$ we denote the (super)symmetric algebra. Its homogeneous components $\Omega^i X$ are locally free sheaves on X, and there is defined the de Rham complex $(\Omega^{\cdot}X, d)$, where $d : \Omega^i X \to \Omega^{i+1} X$ acts by the formula: $d = \sum_i du_i \frac{\partial}{\partial u_i}$, u_i being homogeneous coordinates on X, $u_1, \cdots, u_n$ are even and $u_{n+1}, \cdots, u_{n+m}$ odd. Further if a is an homogeneous element of a $\mathbb{Z}_2$ -graded object, by $\overline{a}$ we denote its parity: $\overline{a}$ is 0 or 1.

The analogue of the canonical sheaf on a supermanifold is the sheaf of volume forms $BerX$, which we call the Berezinian. This is an invertible sheaf of O_x -modules of parity $\frac{1}{2}(1-(-1)^{n+m})$. It is convenient to write its sections in the form:

$$f(\kappa, \xi).D(d\kappa_1 \cdots d\kappa_n d\xi_1 \cdots d\xi_n)$$

where $f(\kappa, \xi)$ is a local section of O_x, $u = (\kappa, \xi)$, $u_i = \kappa_i$, $i \le n$, $u_j = \xi_{j-n}$, $n+1 \le j \le n+m$, [L], [BL1]; D is multiplied by the Berezinian of the Jacobian matrix under the change of coordinates.

Iff $m \ne 0$, $BerX$ does not appear among the Ω^i -s (if $m=0$, $BerX = \Omega^n X$). In this case $BerX$ is included in another complex - the complex of integral forms $\sum_{\cdot} X$, [BL1],

[BL3], the members of which are defined by:

$$\sum_{n-i} X = BerX \otimes_{O_x} \mathbf{S}^i(\prod \mathcal{T}X),\ i \geq 0.$$

Its differential $\partial : \sum_{n-i} X \to \sum_{n-i+1} X$ acts in the following way:

$$\partial(D(du)\otimes Q) = (-1)^{n+m} D(du) \otimes \sum_j \frac{\partial^2 Q}{\partial(\prod \frac{\partial}{\partial u_j})\partial u_j}$$

(here Q is a local section of $\mathbf{S}^i(\prod \mathcal{T}X)$, and $\prod \frac{\partial}{\partial u_j}$ are local sections of $\prod \mathcal{T}X$, $\overline{\prod \frac{\partial}{\partial u_j}} = \overline{\frac{\partial}{\partial u_j}} + 1 = \overline{u_j} + 1$).

We note that d and ∂ are odd k -linear maps.

Iff $m = 0$, there is a canonical isomorphism between $\Omega^{\cdot} X$ and $\sum_{\cdot} X$.

Let ϕ be a vector field. We shall use also the Lie derivative

$$L_\phi : BerX \to BerX$$

defined in [BL3] by the formula $L_\phi = \partial \circ i_\phi$, where i_ϕ is the operator of left tensor multiplication by $\prod \phi$:

$$i_\phi : BerX \to BerX \otimes \prod \mathcal{T}X,$$

$$(\prod \phi = \sum (-1)^{\overline{\phi_i}} \phi_i \prod \frac{\partial}{\partial u_i}, \text{ if } \phi = \sum \phi_i \frac{\partial}{\partial u_i}).$$

§2. Left and Right Connections

2.1. Denote by $\mathcal{D}_{\leq 1}$ the sheaf of differential operators on X of order ≤ 1, and by $\mathcal{F}$ a locally free sheaf of O_x -modules.

Definition. A left connection on $\mathcal{F}$ is an even k -linear map: $\Delta_l : \mathcal{D}_{\leq 1} \otimes_k \mathcal{F} \to \mathcal{F}$, with the conditions:

$$\begin{aligned}
&\text{L0.} \quad \Delta_l(a \otimes f) &&= af \\
&\text{L1.} \quad \Delta_l(\phi \otimes af) &&= \Delta_l(\phi a \otimes f) \\
&\text{L2.} \quad \Delta_l(a\phi \otimes f) &&= a\Delta_l(\phi \otimes f)
\end{aligned}$$

where ϕ is a vector field - a local section of $\mathcal{T}X$, and a is a function - a local section of O_x.

A right connection on $\mathcal{F}$ is an even k -linear map $\Delta_r : \mathcal{F} \otimes_k \mathcal{D}_{\leq 1} \to \mathcal{F}$, with the conditions:

$$\begin{aligned}
&\text{R0.} \quad \Delta_r(f \otimes a) &&= fa \\
&\text{R1.} \quad \Delta_r(f \otimes \phi a) &&= \Delta_r(f \otimes \phi)a \\
&\text{R2.} \quad \Delta_r(fa \otimes \phi) &&= \Delta_r(f \otimes a\phi).
\end{aligned}$$

Note that R1 and L1 imply that connections are not O_x -linear maps, for example:

$$\begin{aligned}
a\Delta_r(f\otimes\phi) &= (-1)^{\overline{a}(\overline{f}+\overline{\phi})}\Delta_r(f\otimes\phi)\cdot a \\
&= (-1)^{\overline{a}(\overline{f}+\overline{\phi})}\Delta_r(f\otimes\phi a) \\
&= (-1)^{\overline{a}(\overline{f}+\overline{\phi})}\Delta_r(f\otimes\phi(a)) + (-1)^{\overline{a}\overline{f}}\Delta_r(f\otimes a\phi) \\
&= \Delta_r(af\otimes\phi) + (-1)^{(\overline{a}+\overline{f})\overline{\phi}}\phi(a)f.
\end{aligned}$$

If we fix a vector field ϕ, we obtain operators $\Delta_l(\phi)$ and $\Delta_r(\phi)$:

$$\begin{aligned}\Delta_l(\phi)(f) &= \Delta_l(\phi\otimes f)\\ \Delta_r(\phi)(f) &= (-1)^{\bar{f}\bar{\phi}}\Delta_r(f\otimes\phi)\end{aligned}$$

which we call as usual covariant derivatives along ϕ.

2.2. The sequences of de Rham and Spencer. For every sheaf $\mathcal{F}$ with a left connection Δ_l we can define its de Rham sequence $DR^{\cdot}(\mathcal{F})$ and for every sheaf $\mathcal{F}$ with right connection Δ_r we can define its Spencer sequence $S_{\cdot}(\mathcal{F})$:

Definition. To a pair $(\mathcal{F},\Delta_l)$ we associate $DR^{\cdot}(\mathcal{F})$, where $DR^i(\mathcal{F})=\Omega^iX\otimes_{O_x}\mathcal{F}$, and for every $i\geq 0$ we have a differential operator $\nabla_l(i):\Omega^iX\otimes_{O_x}\mathcal{F}\to\Omega^{i+1}X\otimes\mathcal{F}$ given by the formula:

$$\nabla_l(i)(du_{K_1}\cdots du_{K_i}\otimes f)=\sum_j(-1)^{\bar{u}_j(\sum_1^i\overline{du_{K_r}})}du_j\cdot du_{K_1}\cdots du_{K_i}\otimes\Delta_l(\frac{\partial}{\partial u_j}\otimes f).$$

Definition. To a pair $(\mathcal{F},\Delta_r)$ we associate its Spencer sequence $S_{\cdot}(\mathcal{F})$, where $S_{n-i}(\mathcal{F})=\mathcal{F}\otimes_{O_x}\mathbf{S}^i(\prod\mathcal{T}X)$ $(\mathbf{S}(\prod\mathcal{T}X)$ is the supersymmetric algebra of $\prod\mathcal{T}X)$, and for every $i>0$ we have a differential operator
$\nabla_r(-i+n)\ :\mathcal{F}\otimes_{O_x}\mathbf{S}^i(\prod\mathcal{T}X)\to\mathcal{F}\otimes_{O_x}\mathbf{S}^{i-1}(\prod\mathcal{T}X)$, given by the formula:

$$\nabla_r(-i+n)(f\otimes\pi\frac{\partial}{\partial u_{l_1}}\cdots\frac{\partial}{\partial u_{l_i}})=$$

$$\sum_k(-1)^{(\sum_{j=1}^{k-1}\bar{u}_{l_j}+k-1)(\bar{u}_{l_k}+1)+\bar{f}+1}\Delta_r(f\otimes\frac{\partial}{\partial u_{l_k}})\otimes(\pi\frac{\partial}{\partial u_{l_1}}\cdots\widehat{\frac{\partial}{\partial u_{l_k}}}\cdots\frac{\partial}{\partial u_{l_i}}).$$

It is easy to check that $\nabla_l(i)$ and $\nabla_r(-i+n)$ are defined invariantly by Δ_l and Δ_r and are odd maps. We call them the i-th covariant differentials of Δ_l or Δ_r. Of course in general $DR^{\cdot}(\mathcal{F})$ and $S_{\cdot}(\mathcal{F})$ are not complexes.

Remark. The connections Δ_l and Δ_r are uniquely determined by their covariant differentials $\nabla_l(1)$ and $\nabla_r(n-1)$ because the action of $\phi\cdot a$ is uniquely determined by the action of $a\phi$ (Leibniz rule), so it is sufficient to give explicitly only the action of vector fields, or equivalently the covariant differentials $\nabla_l(1)$ or $\nabla_r(n-1)$.

2.3. The curvature of a connection. As usual, in order to define the curvature of a connection we have to consider the operators $\nabla_l(i+1)\cdot\nabla_l(i)$ and $\nabla_r(-i+n)\cdot\nabla_r(-i+n-1)$ in more detail.

2.3.1.**Lemma:** The operators $\nabla_l(i+1)\cdot\nabla_l(i)$ and $\nabla_r(-i+n)\cdot\nabla_r(-i+n-1)$ are O_x -linear, $\forall i\geq 0$.

Proof: The statement about $\nabla_l(i+1)\cdot\nabla_l(i)$ may be checked exactly as in the case $m=0$, thus we omit this computation and turn to the case of a right connection Δ_r. Consider first $\nabla_r(n-1)\cdot\nabla_r(n-2):\mathcal{F}\otimes\mathbf{S}^2(\prod\mathcal{T}X)\to\mathcal{F}$. We have:

$$\nabla_r(n-1)\cdot\nabla_r(n-2)(f\otimes\pi\frac{\partial}{\partial u_i}\pi\frac{\partial}{\partial u_j})=$$

$$(-1)^{\bar{u}_i}[\Delta_r(\Delta_r(f\otimes\frac{\partial}{\partial u_i})\otimes\frac{\partial}{\partial u_j})-(-1)^{\bar{u}_i\bar{u}_j}\Delta_r(\Delta_r(f\otimes\frac{\partial}{\partial u_j})\otimes\frac{\partial}{\partial u_i})].$$

The linearity of this map may be checked directly - let us show for example, that:

$$\nabla_r(n-1)\cdot\nabla_r(n-2)(af\otimes\pi\frac{\partial}{\partial u_i}\pi\frac{\partial}{\partial u_j}) = a\nabla_r(n-1)\cdot\nabla_r(n-2)(f\otimes\pi\frac{\partial}{\partial u_i}\pi\frac{\partial}{\partial u_j}).$$

We have :

$$\nabla_r(n-1)\cdot\nabla_r(n-2)(af\otimes\pi\frac{\partial}{\partial u_i}\pi\frac{\partial}{\partial u_j}) =$$

$$= (-1)^{\bar{a}\bar{f}+\bar{u}_i}[\Delta_r(\Delta_r(f\otimes a\frac{\partial}{\partial u_i})\otimes\frac{\partial}{\partial u_j}) - (-1)^{\bar{u}_i\bar{u}_j}\Delta_r(\Delta_r(f\otimes a\frac{\partial}{\partial u_j})\otimes\frac{\partial}{\partial u_i})] =$$

$$= (-1)^{\bar{a}\bar{f}+\bar{u}_i}\{(-1)^{\bar{a}\bar{u}_i}[\Delta_r(\Delta_r(f\otimes\frac{\partial}{\partial u_i}\cdot a)\otimes\frac{\partial}{\partial u_j}) - \Delta_r(f\otimes\frac{\partial}{\partial u_i}(a)\cdot\frac{\partial}{\partial u_j})] -$$

$$-(-1)^{\bar{u}_i\bar{u}_j+\bar{a}\bar{u}_j}[\Delta_r(\Delta_r(f\otimes\frac{\partial}{\partial u_j}\cdot a)\otimes\frac{\partial}{\partial u_i}) - \Delta_r(f\otimes\frac{\partial}{\partial u_j}(a)\cdot\frac{\partial}{\partial u_i})]\} =$$

$$= (-1)^{\bar{a}\bar{f}+\bar{u}_i}\{[(-1)^{\bar{a}\bar{u}_i+\bar{a}\bar{u}_j}\Delta_r(\Delta_r(f\otimes\frac{\partial}{\partial u_i})\otimes\frac{\partial}{\partial u_j}\cdot a) -$$

$$-(-1)^{\bar{a}(\bar{u}_i+\bar{u}_j)}\Delta_r(f\otimes\frac{\partial}{\partial u_i})\frac{\partial}{\partial u_j}(a) - (-1)^{\bar{a}\bar{u}_i+(\bar{u}_i+\bar{a})\bar{u}_j}\Delta_r(f\otimes\frac{\partial}{\partial u_j})\frac{\partial}{\partial u_i}(a) +$$

$$+(-1)^{\bar{a}\bar{u}_i+(\bar{u}_i+\bar{a})\bar{u}_j}f\frac{\partial}{\partial u_j}(\frac{\partial}{\partial u_i}(a))] -$$

$$-(-1)^{\bar{u}_i\bar{u}_j}[(-1)^{\bar{a}\bar{u}_j+\bar{a}\bar{u}_i}\Delta_r(\Delta_r(f\otimes\frac{\partial}{\partial u_j})\frac{\partial}{\partial u_i}\cdot a) -$$

$$-(-1)^{\bar{a}(\bar{u}_i+\bar{u}_j)}\Delta_r(f\otimes\frac{\partial}{\partial u_j})\frac{\partial}{\partial u_i}(a) - (-1)^{\bar{a}(\bar{u}_i+\bar{u}_j)+\bar{u}_i\bar{u}_j}\cdot$$

$$\cdot\Delta_r(f\otimes\frac{\partial}{\partial u_i})\frac{\partial}{\partial u_j}(a) + (-1)^{\bar{a}(\bar{u}_i+\bar{u}_j)+\bar{u}_i\bar{u}_j}f\frac{\partial}{\partial u_i}(\frac{\partial}{\partial u_j}(a))]\} =$$

$$= (-1)^{(\bar{a}\bar{f}+\bar{u}_i+\bar{a}(\bar{u}_i+\bar{u}_j))}[\Delta_r(\Delta_r(f\otimes\frac{\partial}{\partial u_i})\otimes\frac{\partial}{\partial u_j}\cdot a) -$$

$$-(-1)^{\bar{u}_i\bar{u}_j}\Delta_r(\Delta_r(f\otimes\frac{\partial}{\partial u_j})\otimes\frac{\partial}{\partial u_i}\cdot a)] =$$

$$= (-1)^{\bar{a}\bar{f}+\bar{a}(\bar{u}_i+\bar{u}_j)}\nabla_r(n-1)\cdot\nabla_r(n-2)(f\otimes\pi\frac{\partial}{\partial u_i}\pi\frac{\partial}{\partial u_j})\cdot a =$$

$$= a\nabla_r(n-1)\cdot\nabla_r(n-2)(f\otimes\pi\frac{\partial}{\partial u_i}\pi\frac{\partial}{\partial u_j}).$$

We have now an even O_x -linear map:

$$\nabla_r(n-1)\cdot\nabla_r(n-2): \mathcal{F}\otimes\mathbf{S}^2(\prod\mathcal{T}X)\to\mathcal{F}$$

which may be identified with a section of $(End\mathcal{F}\otimes\Omega^2X)_0$. Let's denote it by F_{∇_r}. It is easy to see that all the maps $\nabla_r(-i+n)\cdot\nabla_r(-i+n-1)$ coincide with the inner multiplication by F_{∇_r}. We call F_{∇_r} the curvature of Δ_r.

Considering left connections we get F_{∇_l}, which is a section of $(End\mathcal{F} \otimes \Omega^2 X)_0$ again, but the maps $\nabla_l(i+1) \cdot \nabla_l(i)$ are just multiplication by F_{∇_l} in the algebra $End\mathcal{F} \otimes \Omega^{\cdot} X$. □

2.3.2.**Definition.** A connection is called integrable iff its curvature is zero.

Remark. Let $\mathcal{D}$ be the ring of differential operators on X. It is standard to check that endowing a sheaf $\mathcal{F}$ with a left integrable connection is equivalent of endowing $\mathcal{F}$ with the structure of a left $\mathcal{D}$ -module; the integrability condition allows us to construct a left action of $\mathcal{D}$, coinciding with Δ_l on the differential operators of order ≤ 1. Similarly to give a right connection on $\mathcal{F}$ means to give a right $\mathcal{D}$ -module structure on $\mathcal{F}$.

2.3.3.**Remark.** For pairs $(\mathcal{F}, \Delta_l)$ with $F_{\nabla_l} = 0$, the de Rham sequence $DR^{\cdot}(\mathcal{F})$ is a complex and the Poincaré lemma is true:

$$\begin{aligned} H^i(DR^{\cdot}(\mathcal{F})) &= 0, \qquad i \neq 0 \\ H^0(DR^{\cdot}(\mathcal{F})) &= \mathcal{F}_{hor} \end{aligned}$$

where $\mathcal{F}_{hor}$ is the local system of horizontal sections of $\mathcal{F}$. For the case $(0, d)$ this is proved in [BL3]. The general case is similar. Given a pair $(\mathcal{F}, \Delta_r)$, with Δ_r integrable, the Spencer sequence is by definition also a complex. Here we have the following analogue of the Poincaré lemma:

$$\begin{aligned} H^i(S.(\mathcal{F})) &= 0, \qquad i \neq 0 \\ h^0(S.(\mathcal{F})) &= \mathcal{F}^{hor} \end{aligned}$$

where $\mathcal{F}^{hor}$ is also a local system, which is a subquotient of $\mathcal{F} \otimes \mathbf{S}(\prod \mathcal{T}X)$. We shall explain this later in more detail.

§3. The Equivalence Theorem

In this section we describe a procedure, allowing given a left connection on $\mathcal{F}$, to construct a right connection on $BerX \otimes_{\mathcal{O}_x} \mathcal{F}$. This gives rise to an equivalence of the categories of pairs $(\mathcal{F}, \Delta_l)$ and (G, Δ_r).

3.1. Let us first establish some properties of the complexes of differential and integral forms:

3.1.1.**Proposition.** Let η be a local section of $\Omega^i X$. Then the following Leibniz formulae are true:

$$\begin{aligned} \partial(\eta \cdot s) &= -\partial\eta \cdot s + (-1)^{\overline{\eta}}\eta\partial s \qquad &\text{(means inner multiplication)} \\ d(\eta \cdot \zeta) &= d\eta \cdot \zeta + (-1)^{\overline{\eta}}\eta \cdot d\zeta \qquad &\text{(means multiplication in } \mathbf{S}(\Omega^1 X)) \end{aligned}$$

where s is an integral form, and ζ - a differential form.

The proof is a straightforward local computation.

3.1.2. **Proposition.** The sheaf $BerX$ has a canonical right integrable connection, whose Spencer complex can be identified with $\Sigma.X$.

Proof: We define the map Δ_r by:

$$\Delta_r(\omega \otimes \phi) = -(-1)^{\overline{\omega}\overline{\phi}} L_\phi \omega,$$

where ω is a local section of $BerX$, and L_ϕ - the Lie derivative, see 1.2.

An easy computation shows that Δ_r is really a connection and that $\nabla_r(-i+n) : BerX \otimes \mathbf{S}^i(\prod \mathcal{T}X) \to BerX \otimes \mathbf{S}^{i-1}(\prod \mathcal{T}X)$ is identified with ∂. This means that $BerX$ has a canonical structure of a right $\mathcal{D}$ -module, [P]. □

3.2. The functors B and B^{-1}.

Given a pair $(\mathcal{F}, \Delta_l)$, consider now the pair $(BerX \otimes_{O_x} \mathcal{F}, B_{\Delta_l})$, where B_{Δ_l} acts on $BerX \otimes_{O_x} \mathcal{F}$ as follows:

$$B_{\Delta_l}(\omega \otimes f \otimes \phi) = -((-1)^{\overline{\phi}(\overline{f}+\overline{\omega})} L_\phi \omega \otimes f + (-1)^{\overline{\phi}\,\overline{f}} \omega \otimes \Delta_l(\phi \otimes f)).$$

3.2.1.**Proposition.** a) B_{Δ_l} is a right connection.

b) $F_{B_{\Delta_l}} = 0 \Longleftrightarrow F_{\Delta_l} = 0$.

Proof: This can also be checked directly. We say only that b) means that the functor B transforms left $\mathcal{D}$ -modules into right $\mathcal{D}$ -modules. This statement is well known, cf. [P]□.

3.2.2. The inverse functor B^{-1} transforms right connections on $\mathcal{F}$ into left connections on the sheaf $Hom_{O_x}(BerX, \mathcal{F}) = (BerX)^* \otimes \mathcal{F}$. By definition $b^{-1}\Delta_r$ acts as follows:

$$B^{-1}\Delta_r(\phi \otimes g)(\omega) = (-1)^{\overline{\phi}(\overline{g}+\overline{\omega})}\{g(\Delta_r(\omega \otimes \phi)) - \Delta_r(f(\omega) \otimes \phi)\},$$

(here, g is a local section of $Hom_{O_x}(BerX, \mathcal{F})$).

B^{-1} is really the inverse functor to B. Let us check for example that $B^{-1}(BerX) = Hom(BerX, BerX)$ is just O_x with its canonical left connection $\Delta_l(\phi \otimes f) = \phi(f)$, ($f$ being a local section of $Hom(BerX, BerX) = O_x$). We have:

$$B^{-1}\Delta_r(\phi \otimes f)\omega =$$

$$= (-1)^{\overline{\phi}(\overline{f}+\overline{\omega})}\{f \cdot \Delta_r(\omega \otimes \phi) - \Delta_r(f \cdot \omega \otimes \phi)\} =$$

$$= (-1)^{\overline{\phi}(\overline{f}+\overline{\omega})+\overline{\omega}\overline{\phi}} f \cdot L_\phi \omega + L_\phi(f \cdot \omega) =$$

$$(-1)^{\overline{\phi}\,\overline{f}} f \cdot L_\phi \omega + L_\phi(f) \cdot \omega + (-1)^{\overline{\phi}\,\overline{f}} f \cdot L_\phi(\omega) = \phi(f).$$

The general case is similar. Now one can state:

3.2.3. **Proposition.** B and B^{-1} are inverse equivalences of the categories of pairs $(\mathcal{F}, \Delta_l)$ and (G, Δ_r), in which the morphisms are sheaf morphisms, commuting with the connections.

Remark. The notion of a right connection is of course essential only in the case $m \neq 0$, because for manifolds the Spencer sequence $S_.(\mathcal{F})$ of $(\mathcal{F}, \Delta_r)$ is canonically isomorphic to $DR^\cdot(B^{-1}(\mathcal{F}))$.

We explain now the notion of horizontal sections for a right integrable connection. It is easy to check that for a left integrable connection its sheaf of horizontal sections $\mathcal{F}_{hor}$ is canonically identified with the sheaf of morphisms of O_x into $\mathcal{F}$, as sheaves with left connections or as$\mathcal{D}$ -modules: $\mathcal{F}_{hor} = Hom_{\mathcal{D}}(O_x, \mathcal{F})$. Similarly we can define $\mathcal{F}^{hor} = Hom_{\mathcal{D}}(BerX, \mathcal{F})$, where $\mathcal{F}^{hor}$ is now equipped with a right integrable connection. $\mathcal{F}^{hor}$ is a local system, and $rk\mathcal{F}^{hor} = rk\mathcal{F}$. One may check that $\mathcal{F}^{hor}$ is the only cohomology of the complex of sheaves $S_.(\mathcal{F})$. The equivalence theorem implies that $B(\mathcal{F}^{hor}) = \mathcal{F}_{hor}$.

3.3. Coordinate computations. Now we want to give an explicit coordinate description of left and right connections.

As usual, a left connection Δ_l is determined by its connection form $\chi \in \Gamma(\Omega^1 X \otimes End\mathcal{F})_1$, which is an odd matrix of differential forms (in the even case all matrices from $\Omega^1 X \otimes End\mathcal{F}$ are automatically odd, because $\Omega^1 X$ is odd). The connection form depends, of course, on the choice of a trivialization of $\mathcal{F}$ and acts by the formula

$$\nabla_l(i)(f) = df + \chi f$$

where df means differentiation of the components of f (which is a local section of $\mathcal{F} \otimes \Omega^i X$) in this trivialization. (Note that $\nabla_l(i) = d + \chi$ is odd, as required). Invariantly this means that if the set of all left connections on $\mathcal{F}$ – $Conn_l(\mathcal{F})$ is not empty, then it is a principal homogeneous space over $\Gamma(X, End\mathcal{F} \otimes \Omega^1 X)_1$.

Naturally the set of all right connections on $\mathcal{F}$ – $Conn(\mathcal{F})$ is also either empty or a principal homogeneous space over $\Gamma(X, End\mathcal{F} \otimes \Omega^1 X)_1$, because of the equivalence theorem. (In the category of C^∞ -manifolds, $Conn_l(\mathcal{F})$ and $Conn_r(\mathcal{F})$ are never empty; this can happen only in the analytic category).

Proposition. Let $(\mathcal{F}, \Delta_l)$ have a connection form χ in some trivialization of $\mathcal{F}$. Then

$$B(\nabla_l)(-i+n) : \mathcal{F} \otimes BerX \otimes \mathbf{S}^i(\textstyle\prod \mathcal{T}X) \to \mathcal{F} \otimes BerX \otimes \mathbf{S}^{i-1}(\prod \mathcal{T}X)$$

acts by the formulas $\nabla_r(s) = \partial s - \chi s$, where ∂ acts on the components of s (they are sections of $BerX \otimes \mathbf{S}^i(\prod \mathcal{T}X)$ and χ acts by inner multiplication).

The proof is a direct computation using the formula from 3.2.1.

Consider now the curvature of a left connection. As usual one may compute it in the following way:

$$F_{\nabla_l} = (d+\xi)(d+\xi) = d^2 + \xi \cdot d + d \cdot \xi + \xi \cdot \xi =$$

$$= 0 + \xi \cdot d + d(\xi) + (-1)^{\bar{\xi}\bar{d}} \xi \cdot d + \xi \cdot \xi = d(\xi) + \xi \cdot \xi \ .$$

Similarly one may compute the curvature of the right connection $B(\mathcal{F}, \Delta_l)$:

$$F_{\nabla_r} = (\partial - \xi)(\partial - \xi) = \partial^2 - \xi \cdot \partial - \partial \cdot \xi + \xi \cdot \xi =$$

$$= 0 - \xi \cdot \partial + d(\xi) - (-1)^{\bar{\partial}\bar{\xi}} \xi \cdot d + \xi \cdot \xi = d(\xi) + \xi \cdot \xi$$

(we use here Leibniz rule from 3.1. and the last proposition).

Corollary: $F_{\nabla_l} = F_{B(\nabla_l)}$.□

3.4. Tensor operations. There are several canonical tensor operations on left connections, for example, if $\mathcal{F}_1$ and $\mathcal{F}_2$ have left connections Δ_{l1} and Δ_{l2}, then Δ_{l1} and Δ_{l2} induce the left connections $\Delta_{l1} \otimes \Delta_{l2}$ on $\mathcal{F}_1 \otimes_{O_x} \mathcal{F}_2$ and $\Delta_{l1}^* \otimes \Delta_{l2}$ on $Hom(\mathcal{F}_1, \mathcal{F}_2)$:

$$\Delta_{l1} \otimes \Delta_{l2}(\phi \otimes f_1 \otimes f_2) = \Delta_{l1}(\phi \otimes f_1) \otimes f_2 + (-1)^{\bar{f_1}\bar{f_2}} \Delta_{l2}(\phi \otimes f_2) \otimes f_1$$

$$\Delta_{l1}^* \otimes \Delta_{l2}(\phi \otimes \tilde{f})(f_1) = \Delta_{l2}(\phi \otimes \tilde{f}(f_1)) - (-1)^{\bar{\tilde{f}}\bar{\phi}} \tilde{f}(\Delta_{l1}(\phi \otimes f_1))$$

where f_i are local sections on $\mathcal{F}_i$ and $\tilde{f}$ is a local section on $Hom(\mathcal{F}_1, \mathcal{F}_2)$.

The tensor operations on right connections are not so evident, for example, if $\mathcal{F}_1$ and $\mathcal{F}_2$ have right connections in the holomorphic category it can happen that the sets $Conn_r(\mathcal{F}_1 \otimes \mathcal{F}_2)$ and $Conn_r(\mathcal{F}_1^* \otimes \mathcal{F}_2)$ are empty. Indeed, let X -be the projective space $\mathbf{P}^n$ and let $\mathcal{F}_1 = \mathcal{F}_2 = BerX = \Omega^n X$ (the sheaf of holomorphic n -forms). Then it is standard to check that the sheaves $(\Omega^n X)^2$ and O_x have no right holomorphic connections at all. The reason of this "asymmetry" is that the sheaf O_x, over which we tensor, has a canonical left connection, but no canonical right connection.

The right way of obtaining correct tensor operations on right connections is to transform them into left connections using the functor B^{-1}, apply there the already known tensor operations and then "come back" by the functor B. For example given two pairs $(\mathcal{F}_1, \Delta_{r1}), (\mathcal{F}_2, \Delta_{r2})$ the sheaf $\mathcal{F}_1 \otimes \mathcal{F}_2 \otimes (BerX)^*$ has a canonical right connection.

3.5. Bianchi identities.

3.5.1. **Proposition.** (Bianchi identity for left connections). Let $\mathcal{F}$ be equipped with a left connection Δ_l. Denote by $\tilde{\Delta}_l$ the corresponding left connection on $End\mathcal{F}$. Then

$$\tilde{\nabla}_l(2)(F_{\nabla_l}) = 0.$$

Exactly the same computation as in the case $m = 0$ proves this statement.

3.5.2. Let $\mathcal{F}$ be now equipped with a right connection Δ_r. The sheaf $End\mathcal{F} \otimes (BerX)^*$ is equipped with the connection $\tilde{\Delta}_r$, where the pair $(End\mathcal{F} \otimes (BerX)^*, \tilde{\Delta}_r)$ is by definition equal to the pair $B(End\mathcal{F}', \widetilde{B^{-1}\Delta_r})$,

$$\mathcal{F}' \otimes BerX = \mathcal{F},$$

and its Spencer sequence has the form:

$$\to End\mathcal{F}' \otimes BerX \otimes \mathbf{S}^n(\textstyle\prod \mathcal{T}X) \to \cdots \to End\mathcal{F}' \otimes BerX \otimes \mathbf{S}^{n-2}(\textstyle\prod \mathcal{T}X) \to \quad .$$

Consider also the complex of integral forms $S.(BerX)$. In $S_0(BerX) = BerX \otimes \mathbf{S}^n(\prod \mathcal{T}X)$ we have a canonical (up to a constant) cohomology class, corresponding to $(BerX)^{hor}$. $(BerX)^{hor}$ is equal to $(B(O_x))^{hor}$ which is the constant sheaf k. Let us denote this class with η. Then $\tilde{\eta} = id \otimes \eta$ lies in $End\mathcal{F} \otimes BerX \otimes \mathbf{S}^n(\prod \mathcal{T}X)$ and we can apply to $F_{\nabla_r} \cdot \tilde{\eta}$ the operator $\tilde{\Delta}_r$.

Proposition (Bianchi identity for right connections):

$$\tilde{\nabla}_r(2)(\mathcal{F}_{\nabla_r} \cdot \tilde{\eta}) = 0.$$

Proof: Let θ be the connection form of $B^{-1}\Delta_r$ in some trivialization of $(BerX)^* \otimes \mathcal{F}$. Then $\tilde{\nabla}_r(\kappa)$ acts as $d + ad\theta$, where $ad\theta(\xi) = \theta\xi - (-1)^{\overline{\theta\xi}}\xi\theta$. We have:

$$(\partial - ad\theta)(F_{\nabla_r} \cdot \tilde{\eta}) = (\partial - ad\theta)(F_{B^{-1}(\nabla_r)} \cdot \tilde{\eta}) =$$

$$= -dF_{B^{-1}(\nabla_r)} \cdot \tilde{\eta} + F_{B^{-1}(\nabla_r)}\partial(\tilde{\eta}) - ad\theta(F_{B^{-1}(\nabla_r)}) \cdot \tilde{\eta} =$$

$$= -\widetilde{B^{-1}(\nabla_r)}(2)[F_{B^{-1}(\nabla_r)}] \cdot \eta + F_{B^{-1}(\nabla_r)} \cdot \partial(\tilde{\eta}).$$

But the first term is equal to zero by the Bianchi identity for left connections, and the second term is zero because $\partial(\tilde{\eta}) = \mathbf{0}$ by definition. □

§4. Remarks on Characteristic classes

If φ is an even homogeneous invariant polynomial on the space of $p|q$ −matrices over a supercommutative ring A−

$$\varphi : M_{p|q}(A) \to A, \qquad \varphi(ZYZ^{-1}) = \varphi(Y),\ \overline{Z} = 0,$$

considering a locally free sheaf $\mathcal{F}, rk\mathcal{F} = p \mid q$, with (smooth) left or right connection Δ_l or Δ_r, we can apply φ to F_{∇_l} or F_{∇_r}, which are sections of $(End\mathcal{F} \otimes \Omega^2 X)_0$ and obtain

$\varphi(F_{\nabla_l})$ and $\varphi(F_{\nabla_r})$ (they are correctly defined global differential forms of degree $2deg\varphi$). As in case $m = 0$, we have the following fundamental result:

Theorem (Chern-Weil)

A) $d\varphi(F_{\nabla_l}) = 0$; $\quad \partial(\varphi(F_{\nabla_r}) \cdot \eta) = 0$.

B) The cohomology classes $\varphi_l \in H^{2deg\varphi}(\Gamma(X, \Omega^{\cdot} X))$ (corresponding to $\varphi(F_{\nabla_l})$) and $\varphi_r \in H^{2deg\varphi}(\Gamma(X, \Sigma_{\cdot} X))$ (coresponding to $\varphi(F_{\nabla_r})$) do not depend on the choice of Δ_l or Δ_r.

The statements about left connections can be proved exactly as in the case $m = 0$, using minimum supercommutative algebra, and we turn to the statements about right connections:

A) $\partial(\varphi(F_{\nabla_r}) \cdot \eta) = \partial(\varphi(F_{B^{-1}(\nabla_r)} \cdot \eta) = -(d\varphi(F_{B^{-1}(\nabla_r)})) \cdot \eta + \varphi(F_{B^{-1}(\nabla_r)}) \cdot \partial\eta = 0$.

The first term is zero because $B^{-1}(\nabla_r)$ is a left connection, and the second term is zero because $\partial\eta = 0$.

B) Let Δ_r' and Δ_r'' be two right connections. We have:

$$\varphi(F_{\nabla_r'}) \cdot \eta - \varphi(F_{\nabla_r''}) \cdot \eta = (\varphi(F_{B^{-1}(\nabla_r')}) - \varphi(F_{B^{-1}(\nabla_r'')})) \cdot \eta =$$

$$= (d\alpha) \cdot \eta = \partial(-\alpha \cdot \eta).$$

Here we use again the fact, that $\partial\eta = 0$, and that $\varphi(F_{B^{-1}(\nabla_r')}) - \varphi(F_{B^{-1}(\nabla_r'')}) = d\alpha$, where α is a differential form.

This allows us to construct the Chern classes of a locally free sheaf $\mathcal{F}$ using the invariant forms $\varphi_r(F_\nabla) = str(F_\nabla^r)$ by two ways. Unfortunately these do not give new interesting invariants of $\mathcal{F}$, because the previous construction uses smooth connections, and it is well known that in the C^∞ -case the category of locally free sheaves on X is equivalent to the category of $\mathbf{Z}_2$ -graded locally free sheaves on the reduced manifold X_{red}. This is a variant of the theorem of Batchelor [B], but may be proved directly too. Thus, the classes we get by the Chern-Weil construction are nothing but the Chern classes of the sheaf $\mathcal{F}_{red}$ (respectively $(\Omega^n X_{red})^* \otimes \mathcal{F}_{red}$).

Finally, we note that the Atiyah-style classes in the holomorphic situation are more interesting because the cohomology ring $\oplus_i H^i(X, \Omega^i X)$ even in the simplest case of $\mathbf{P}^{1|2}$ (the projectivization of a linear space of dimension $2 \mid 2$) does not coincide with the ring $\oplus_i H^i(X_{red}, \Omega^i X_{red})$.

References

[1] [B] Batchelor M., The structure of supermanifolds, Transactions of the AMS, 253, 329-338 (1979).

[2] [BC] Bott R., Chern S. S., Hermitean vector bundles and the equidistribution of the zeroes of their holomorphic sectionsm, Acta Math., 114, 71-112 (1965).

[3] [Bj] Björk J. E., Rings of differential operators, North-Holland, Amsterdam. Oxford. New York, 375p. (1979).

[4] [BL1] Bernstein J. N., Leites D. A., Integrable forms and the Stoke's formula on supermanifolds, Funct. Anal. Appl., 11, 45 (1977).

[5] [BL2] Bernstein J. N., Leites D. A., Integration of differential forms on supermanifolds, Funct. Anal. Appl., **11**, 45 (1977).

[6] [BL3] Bernstein J. N., Leites D. A., Invariant differential operators and irreducible representations of Lie superalgebras of vector fields, Serdika, 7, 320-384, (1981) (in Russian).

[7] [L] Leites D. A., Introduction to the theory of supermanifolds, Russian Mathematical Surveys, **35**, 1 (1980).

[8] [MS] Milnor J., Stasheff J., Characteristic classes, Annals of mathematical studies, **76**, Princeton, New Jersey, (1974).

[9] [P] Penkov I., $\mathcal{D}$ -modules on supermanifolds, Inv. Math., **71**, 501-512, (1983).

[10] [W] Wells R. O., Differential analysis on complex manifolds, Prentice-Hall Inc, Englewood Cliffs, N. J., (1973).

New Lie Superalgebras and Mechanics - Some of the Unmissed Opportunities of the Supermanifold Theory -

D.A.Leites
Karelian Branch of Academy of Sci., USSR, Petrogavodsk

V.V.Minachin
URVŠ, Konventna 1
801 00 Bratislava, Czechoslovakia

In his work [1], which has already become classical, Kac listed four series of simple infinite-dimensional Lie superalgebras. The first of them is

$$W(m,n) = Der\mathbb{C}[[u]] \qquad (I)$$

which consists of derivations of the commutative superalgebra $\mathbb{C}[[u_1, \cdots, u_{m+n}]]$ of formal power series in m even $u_1, \cdots, u_m$ and n odd $u_{m+1}, \cdots, u_{m+n}$variables. Elements of the Lie superalgebra $W(m,n)$ are called vector fields. Any vector field $D \in W(m,n)$ is of the form

$$D = \sum_{i=1}^{m+n} f_i(u)\frac{\partial}{\partial u_i} \qquad , f_i \in \mathbb{C}[[u]].$$

A divergence of the vector field $D \in W(m,n)$ is given by the formula

$$divD = \sum_{i=1}^{m} \frac{\partial f_i}{\partial u_i} - (-1)^{\tilde{D}} \sum_{i=m+1}^{m+n} \frac{\partial f_i}{\partial u_i} \ .$$

The second Kac's series consists of divergence-free vector fields

$$S(m,n) = \{D \in W(m,n) \mid divD = 0\} \ . \qquad (II)$$

W(m,n)-module of differential forms. Consider another set of $m+n$ formal variables $du_1, \cdots, du_{m+n}$ endowed with parity in the following way: $\tilde{du_i} = \tilde{u_i} + 1$. The superalgebra of differential forms $\Omega(m,n)$ is by definition $\Omega(m,n) = \mathbb{C}[[u]][du]$, i.e. polynomials in $du_1, \cdots, du_{m+n}$ with coefficients coming from $\mathbb{C}[[u]]$. There is a natural Z-grading in $\Omega(m,n)$:

$$\Omega(m,n) = \Omega^0 \oplus \Omega^1 \oplus \cdots \oplus \Omega^i \oplus \cdots$$

and three operators similar to those one has to deal with in the theory of usual differential forms:

a: the differential $\quad d : \Omega^i \longrightarrow \Omega^{i+1}; d = \sum_{i+1}^{m+n} du_i \frac{\partial}{\partial u_i}$

b: the inner product $i_D : \Omega^i \longrightarrow \Omega^{i-1}; i_D = \sum_{i=1}^{m+n} (-1)^{\bar{f}_i} f_i \frac{\partial}{\partial(du_i)}$

c: the Lie derivative $L_D = [d, i_D]$.

One can check without much difficulty that $[L_D, L_{D'}] = L_{[D,D']}$ which means that $\Omega(m,n)$ constitutes a W(n,m)-module.

Hamiltonian and contact vector fields. Let us also denote $u_1, \cdots, u_m$ by $x_1, \cdots, x_m$ and $u_{m+1}, \cdots, u_{m+n}$ by $\xi_1, \cdots, \xi_n$. Consider now two specific differential forms

$$h = 2\sum_{i=1}^{k} dx_i dx_{k+i} + \sum_{i=1}^{n} (d\xi_i)^2 \qquad \in \Omega(2k, n)$$

and

$$c = dx_{2k+1} + \sum_{i=1}^{k} (x_i dx_{k+i} - x_{k+i} dx_i) + \sum_{i=1}^{n} \xi_i d\xi_i \qquad \in \Omega(2k+1, n).$$

Now by definition the Hamiltonian vector fields are

$$H(2k, n) = \{D \in W(2k, n) \mid L_D h = 0\} \qquad (III)$$

and the contact vector fields

$$K(2k+1, n) = \{D \in W(2k+1, n) \mid L_D c = fc, f \in \mathbb{C}[[u]]\}\ . \qquad (IV)$$

Explicitly, $H(2k,n)$ consists of vector fields of the form

$$D_f = \sum_{i=1}^{k} (\frac{\partial f}{\partial x_i} \frac{\partial}{\partial x_{k+i}} - \frac{\partial f}{\partial x_{k+i}} \frac{\partial}{\partial x_i}) - (-1)^{\bar{f}} \sum_{i=1}^{n} \frac{\partial f}{\partial \xi_i} \frac{\partial}{\partial \xi_i}; \qquad \text{for } f \in \mathbb{C}[[u]].$$

It can be shown that $[D_f, D_g] = D_{\{f,g\}_{P.B.}}$ where

$$\{f, g\}_{P.B.} = \sum_{i=1}^{k} (\frac{\partial f}{\partial x_i} \frac{\partial g}{\partial x_{k+i}} - \frac{\partial f}{\partial x_{k+i}} \frac{\partial g}{\partial x_i}) - (-1)^{\bar{f}} \sum_{i=1}^{n} \frac{\partial f}{\partial \xi_i} \frac{\partial g}{\partial \xi_i}$$

is called a Poisson bracket. There is a similar way to represent vector fields from $K(2k+1, n)$ and the corresponding contact bracket is

$$\{f, g\}_{K.B.} = \Delta f \frac{\partial g}{\partial x_{2k+1}} - \frac{\partial f}{\partial x_{2k+1}} \Delta g - \{f, g\}_{P.B.}$$

where

$$\Delta f = 2f - \sum x_i \frac{\partial f}{\partial x_i} - \sum \xi_i \frac{\partial f}{\partial \xi_i}.$$

The Darboux theorem. A conjecture [1] that there are no other simple infinite - dimensional Lie superalgebras (under certain very natural conditions) proved to be wrong [3],[6],[7]. First two more series have been found by Leites [3] and much more see in[6],[7]. Probably the reason that they have been missed in [1] stems from the assumption that there is only one canonical 2-form, namely h, which is evidently even. However there is no need for a 2-form to be even (cf. [2]) as is shown by the super version of the Darboux theorem formulated and proved by Leites [3].

Theorem. Let ω be a homogeneous nonsingular closed 2-form on a real supermanifold M. If ω is even then dim $M = 2m \mid n$ and it is possible to choose local coordinates on M such that $\omega = 2\sum_{i=1}^{m} dx_i dx_{m+i} + \sum_{i=1}^{n} \epsilon_i (d\xi_i)^2, \epsilon_i = \pm 1$. If ω is odd then dim$M = n \mid n$ and it is possible to choose local coordinates on M such that $\omega = \sum_{i=1}^{n} dx_i d\xi_i$. We shall denote this last form by l.

L and M series of infinite-dimensional Lie superalgebras.

Denote by

$$L(n) = \{D \in W(n,n) \mid L_D l = 0\}.$$

One can show that any vector field from $L(n)$ is of the form

$$D_f = -\sum_{i=1}^{n}\left(\frac{\partial f}{\partial x_i}\frac{\partial}{\partial \xi_i} + (-1)^{\tilde f}\frac{\partial f}{\partial \xi_i}\frac{\partial}{\partial x_i}\right), \qquad f \in \mathbb{C}[[u]]$$

and another similarity to the Hamiltonian series is that

$$[D_f, D_g] = D_{\{f,g\}_{B.B.}}$$

where $\{f,g\}_{B.B.}$ is the Buttin bracket defined as

$$\{f,g\}_{B.B.} = -\sum_{i=1}^{n}\left(\frac{\partial f}{\partial x_i}\frac{\partial g}{\partial \xi_i} + (-1)^{\tilde f}\frac{\partial f}{\partial \xi_i}\frac{\partial g}{\partial x_i}\right).$$

Finally, consider the 1-form $\alpha = d\xi_{n+1} + \sum_{i=1}^{n}(x_i d\xi_i + \xi_i dx_i) \in \Omega(n, n+1)$ (for which $d\alpha = 2l$) and denote by

$$M(n) = \{D \in W(n,n+1) \mid D\alpha = f\cdot\alpha,\ f \in \mathbb{C}[[u]]\}.$$

Again there is a corresponding bracket - the odd counterpart to the contact bracket

$$\{f,g\}_{M.B.} = \Delta f \frac{\partial g}{\partial \xi_{n+1}} - \frac{\partial f}{\partial \xi_{n+1}}\Delta g + (-1)^{\tilde f \tilde g}\{f,g\}_{B.B.}$$

where

$$\Delta f = 2f - \sum_{i=1}^{2n} u_i \frac{\partial f}{\partial u_i}.$$

Vector fields on supermanifolds. The fundamental theorem in the case of usual even manifolds states that for any nonsingular vector field it is possible to choose local coordinates in such a way that it is simply $\frac{\partial}{\partial x_1}$. To formulate a super version of that theorem one needs a coherent notion of a nonsingular vector field in the super case. There was a beautiful idea by Shander [5] who realized that in the usual even case nonsingularity of the vector field D (i.e., that $D(x_0) \neq 0$) is equivalent to the fact that for some neighbourhood U of x_0 the mapping $D\mid_U: C^\infty(u) \longrightarrow C^\infty(u)$ is an epimorphism. It turns out that precisely this formulation can be taken as a definition of a nonsingular vector field in the super case.

Theorem. (Shander [5]). Let D be a vector field on a supermanifold M which is nonsingular at a point $m_0 \in M$. Then it is possible to choose local coordinates on M around m_0 such that

$$D = \frac{\partial}{\partial x_1} \text{if } D \text{ is even, and}$$

$$D = \frac{\partial}{\partial \xi_1} + \xi_1 \frac{\partial}{\partial x_1} \text{if } D \text{ is odd.}$$

This theorem gives rise to the two alternative ways to formulate the notion of an ordinary differential equation on the supermanifold.

Let $(\mathcal{T}_0 \subset \mathcal{T}, u, D_{\mathcal{T}})$ where $\mathcal{T}_0$ is a superdomain in u - coordinates on $\mathcal{T}_0$ and $D_{\mathcal{T}}$ a vector field on M which is of one of the following forms:

a) $(I^{1|0} \subset \mathbb{R}^{1|0}, t, \frac{\partial}{\partial t})$ where $I^{1|0}$ is an interval containing zero.

b) $(I^{1|1} \subset \mathbb{R}^{1|1}, (t, \tau), \frac{\partial}{\partial \tau} + \tau \frac{\partial}{\partial t})$ where $I^{1|1} = I^{1|0} \times \mathbb{R}^{0,1}$.

The supermanifold $\mathcal{T}$ is called the time -supermanifold and $D_{\mathcal{T}}$ - differentiation by the time $\mathcal{T}$.

It turns out then that all the existence and uniqueness theorems on the usual ordinary differential equations can be stated and proved in the super case.

Classical mechanics. By a classical mechanical system in the usual case we mean a symplectic manifold (M, ω) with a distinguished function H on it - a Hamiltonian function. The dynamics of such a system is given by the Poisson bracket $\frac{\partial f}{\partial t} = \{f, H\}_{P.B.}$, where f, H both depend also on t. Now that we have the odd counterpart to the Poisson bracket and the odd counterpart of the usual differentiation by time one easily sees that in the super case two different kind of mechanics are possible. They are given by the formula

$$D_{\mathcal{T}} f = \{f, H\}_{\omega}$$

where the time-supermanifold $\mathcal{T}$, the 2-form ω and the Hamiltonian H should be chosen in such a way that $\tilde{H} + \tilde{\omega} = \tilde{D}_{\mathcal{T}}$.

Recently, the odd mechanics was applied with success to quantization of gauge fields [8].

References

[1] V.Kac, Adv. Math. **26** (1977).

[2] B.Kostant, Lecture Notes in Math. **570**, (1977)

[3] D.Leites, Sov. Math. Doklady **18** N 3, (1977).

[4] D.Leites, Funct. Anal. Appl. **15** N 1, (1982).

[5] V.Shander, Funct.Anal. Appl. **14** N 2 (1980).

[6] D.V. Alekseevrski, D.A. Leites, I.M. Shchepochkina, C.R de l'Acad. bulgare des Sci, **33** N 9, 1187 - 1190, (1980) (in Russian).

[7] I.M. Shchepochkina, C.R de l'Acad. bulgare des Sci. (to appear).

[8] I.A. Batalin, G.A. Vilkoviski, Phys. Lett. **102 B** N 1, 27 - 31, (1981)

N = 1 Supergravity from a Geometrical Point of View

E. Sokatchev
INRNE, Sofia, Bulgaria and
JIRN, Dubna, USSR

1 Introduction

There are several versions of $N = 1$ supergravity: minimal [1] (12+12 fields), non-minimal [2, 3] (20+20 fields), new minimal [4] (12+12) and, finally, the flexible (28+28) one proposed here. They differ in their auxiliary field sets as well as in their intrinsic geometries [5]. In the absence of matter the auxiliary fields vanish on-shell, and it is clear that all the versions are equivalent. In the presence of matter, however, the auxiliary fields are expressed in terms of the matter fields in various ways. Does this lead to different effective interactions? Are there essentially different mechanisms of supersymmetry breaking in those versions of supergravity?

These questions are intensively discussed [6, 7, 8] now in connection with the possible applications of $N = 1$ supergravity in the phenomenological models of grand unification (see [9] and references therein).

The present talk is also devoted to this problem. The geometric approach [5, 10] (see also [3]) developed earlier allows to discuss and compare the different versions of $N = 1$ supergravity on a common basis. The main results are listed here.

i) The most restrictive version of $N = 1$ supergravity is the new minimal one. It demands R-symmetry in matter couplings because of its local $U(1)$ invariance (or, equivalently, because of a rigid constraint on the supergravity prepotentials [5]). We propose here a new version with 28+28 fields which we shall refer to as the flexible one. it is obtained by relaxing the new minimal one. In other words, a Lagrange multiplier is introduced producing the above mentioned constraint for the new minimal version when there is no matter [5]. In the presence of matter, however, the Lagrange multiplier appears in the matter sector as well. This leads to a modified ("self-adjusting") constraint the form of which is influenced by matter couplings. This explains why the flexible version allows the same types of matter couplings as the minimal and non-minimal ones. Notice that the Lagrange multiplier introduced is at the same time a gauge compensator for the local $U(1)$ symmetry of the new minimal version. This is another explanation of the increased versatility of the flexible version.

ii) It has recently been shown that the old and new minimal and the non-minimal versions are mutually equivalent for R-symmetric matter Lagrangians [6]. However, the importance of R-non-invariant matter couplings is evident because the R-symmetry can be broken by anomalies in the quantum case. At the same time there is a common belief [6, 11, 12] that in the non-minimal supergravity R-non-invariant matter couplings are imposible because of the absence of a proper chiral density for the superpotentials despite such a density has been mentioned several years ago [3]. Here we rederive a density of this type and obtain the corresponding superfield equations of motion for a general R-non-invariant superpotential. The consistency of these equations is proved. The auxiliary fields do not propagate and are expressible as combinations of matter fields in a Lorenz-invariant manner. So, the non-minimal version is not more restrictive in matter couplings than the minimal one.

iii) In the flexible supergravity a similar chiral density can be constructed with the help of the Lagrange multiplier (just therefore the matter fields become involved in the constraint mentioned above). Therefore the flexible version also admits general R-non-invariant matter couplings.

It is worthwhile to make the following comment concerning both statements ii) and iii). The chiral densities obtained have some superfield in the denominator. Then one must have a non-vanishing supercosmological term. At the same time the cosmological term in x-space can vanish because it obtains also a contribution opposite in sign from a spontaneous supersymmetry breaking term in the matter sector [13].

2 Geometry of $N = 1$ supergravity

Here we shall briefly recall some necessary information about the geometric formulations of the various $N = 1$ supergravity models. More details can be found in Refs. [5], [10].

Consider the complex superspace

$$\begin{aligned} \mathbb{C}^{414} \;=\; & \{\, Z_L^M = (x_L^m \,,\, \theta_L^\mu \,,\, \bar{\varphi}_L^{\dot\mu})\} \qquad or \\ & \{\, Z_R^M = (x_R^m \,,\, \bar{\theta}_R^{\dot\mu} \,,\, \varphi_R^\mu)\} \;\;;\;\; Z_R^M = (Z_L^M)^+ \;\;. \end{aligned} \tag{1}$$

Here Z_L^M and Z_R^M are left- or right-handed parametrizations of $\mathbb{C}^{414}$. Further, introduce the following "triangular" gauge group:

$$\begin{aligned} \delta\, x_L^m &= \lambda^m (\, x_L, \theta_L \,) \;, \\ \delta\, \theta_L^\mu &= \lambda^\mu (\, x_L, \theta_L \,) \;, \\ \delta\, \bar{\varphi}_L^{\dot\mu} &= \bar{\varrho}^{\dot\mu} (\, x_L, \theta_L, \bar{\varphi}_L \,) \;. \end{aligned} \tag{2}$$

It leaves invariant the "chiral" superspace
$\mathbb{C}^{414} / \mathbb{C}^{012} \sim \mathbb{C}^{412}$:

$$\begin{aligned} \mathbb{C}^{412} \;=\; & \{\, \zeta_L^M = (\, x_L^m, \theta_L^\mu \,)\} \qquad or \\ & \{\zeta_R^M = (\, x_R^m, \bar{\theta}_R^{\dot\mu} \,)\} \;;\; \zeta_R^M = (\, \zeta_L^M \,)^+ \;. \end{aligned} \tag{3}$$

The real (physical) superspace is defined as a hypersurface in $\mathbb{C}^{414}$:

$$\mathbb{R}^{414} = \{\, Z^M = (\, x^m \,,\, \theta^\mu \,,\, \bar{\theta}^{\dot\mu} \,)\} \tag{4}$$

where

$$\begin{aligned} x^m = Re\, x_L^m \quad , \quad \theta^\mu = \theta_L^\mu \;\;,\;\; \bar{\theta}^{\dot\mu} &= \bar{\theta}_R^{\dot\mu} \;; \\ Im\, x_L^m &= H^m (\, x \,,\, \theta \,,\, \bar{\theta} \,) \;, \\ \varphi_R^\mu - \theta_L^\mu &= H^\mu (\, x \,,\, \theta \,,\, \bar{\theta} \,) \;, \\ \bar{\varphi}_L^{\dot\mu} - \bar{\theta}_R^{\dot\mu} &= \bar{H}^{\dot\mu} (\, x \,,\, \theta \,,\, \bar{\theta} \,) \;. \end{aligned} \tag{5}$$

The superfunctions H^m, H^μ, $\overline{H}^{\dot\mu}$ determine the shape of the hypersurface. After imposing a dynamical postulate (action principle) they become the supergravity superfields (prepotentials).

The group (2) corresponds to conformal supergravity. Using superdeterminants (Berezinians) one can impose the following relations between the transformations of the volume elements of $\mathbb{C}^{414}$ and $\mathbb{C}^{412}$:

$$[\, Ber\,(\frac{\partial Z_L'}{\partial Z_L})\,]^{3n+1} = [\, Ber\,(\frac{\partial \zeta_L'}{\partial \zeta_L})\,]^{2n} \quad . \tag{6}$$

These define subgroups of (2) corresponding to the various $N = 1$ Einstein supergravities [1-4]. Depending on the value of the Gates-Siegel paramter n [3] one can distinguish the following three cases:

i) $n = -1/3$. According to (6) in this case

$$Ber\,(\frac{\partial \zeta_L'}{\partial \zeta_L}) = 1 \tag{7}$$

i.e. the supervolume of $\mathbb{C}^{412}$ is preserved. The parameter $\overline{\varrho}^{\dot\mu}$ in (2) remains unrestricted, so the potentials H^μ, $\overline{H}^{\dot\mu}$ can be gauged away, and one is left with H^m only. This is the case of minimal supergravity (12 + 12 fields) [1, 3, 10].

ii) $n = 0$. Now

$$Ber\,(\frac{\partial Z_L'}{\partial Z_L}) = 1 \tag{8}$$

i.e. the supervolume of $\mathbb{C}^{414}$ is preserved. In this case there remains local $U(1)$ invariance in the Wess-Zumino gauge [5, 14]. It causes problems when trying to write down an ivariant action. The first way to deal with this problem is to impose a constraint on the prepotentials H^m, H^μ, $\overline{H}^{\dot\mu}$. It has a clear geometrical meaning. In the case $n = 0$ (and only in it) the Berezinian of changing variables from the left to right-handed parametrization of $\mathbb{C}^{414}$ (see (1), (5))

$$U = Ber\,(\frac{\partial Z_L}{\partial Z_R}) \tag{9}$$

is invariant. Indeed, d^8Z_L and d^8Z_R are invariant (see (8)), and on the hypersurface relates them to each other, $d^8Z_L = U d^8 Z_R$. The constraint on U is

$$U = 1 \quad . \tag{10}$$

It reduces the number of fields to 12+12 again and leads to the so-called new minimal supergravity [4]. Notice that the local $U(1)$ invariance remains in this version of the $n = 0$ theory.

As shown in Ref. [5], there is an alternative approach to the case $n = 0$. The local $U(1)$ invariance can be compensated by introducing a real pseudoscalar compensating superfield $\varphi(x, \theta, \overline{\Theta})$. It transforms as follows:

$$\varphi' = \varphi + \frac{i}{2}(l - r) \tag{11}$$

(The chiral superfunctions-parameters l , r are defined in (18); the first component of $\frac{i}{2}(l-r)$ is just the $U(1)$ parameter). Thus, the number of fields becomes 28+28 (20+20 in H^m, H^μ, $\overline{H}^{\dot\mu}$, 8+8 in φ), no constraints are imposed, no local $U(1)$ invariance is present. This is a new, "flexible" version of the "new minimal" supergravity (or rather, the latter is a truncated version of the former). As we shall see in what follows, this version is more versatile in matter couplings than the new minimal one. In other words, the flexible formulation restores the equal rights of the new minimal version as a member of the family of non-minimal supergravities.

iii) $n \neq -1/3$. 0. In this case we have the 20+20 fields of the so called non-minimal supergravities [2, 3].

Here we would like to make the following comment. Another widely used scheme of classification of $N = 1$ supergravities is the one based on compensating conformal supergravity with various multiplets [15,3]. Thus, the minimal supergarvity corresponds to a chiral compensator, the non-minimal version - to a complex linear one, the new minimal version - to a real linear one. As is shown in [16], the flexible supergravity also belongs to this scheme with a relaxed linear multiplet [17] as a compensator.

3 Building blocks

The standart way of describing the invariant properties of the curved superspace $\mathbb{R}^{414}$ is to develop the formalism of differential geometry, i.e. to introduce supervierbeins, connections, covariant derivatives, torsion, and curvature. All this can be done in the scheme described in section 2. [10]. However, for a number of practical purposes, such as writing down actions, one can avoid using the whole machinery. Instead, one can introduce a few objects with simple transformation properties ("blocks") and then construct various quantities out of them.

We begin defining the basic differential operator in the framework of sect. 2. It is the spinor derivative [5]

$$\nabla_\alpha \Phi(Z) \equiv (1+\Delta H)_\alpha^{-1\beta}\, \Delta_\beta\, \Phi = \frac{\partial}{\partial \varphi_R^\alpha}\, \Phi(Z) \tag{12}$$

of a scalar superfield $\Phi(Z)$ with respect to φ_R^α (in the right-handed parametrization $x_R^m = x^m - i\,H^m$, $\overline{\theta}_R^{\dot\mu} = \overline{\theta}^{\dot\mu}$, $\varphi_R^\mu = \theta^\mu + H^\mu$ of $\mathbb{R}^{414}$, see (5)). Here

$$\begin{aligned} \Delta_\alpha &= \frac{\partial}{\partial\theta^\alpha} - i\,\frac{\partial}{\partial\theta^\alpha}\, H^m\,(1+i\,\partial H)_m^{-1n}\,\frac{\partial}{\partial x^n} \quad , \\ (1+i\,\partial H)_m^n &\equiv \delta_m^n + i\,\partial_m\, H^n \quad , \quad (1+\Delta\, H)_\alpha^\beta \equiv \delta_\alpha^\beta + \Delta_\alpha\, H^\beta \quad . \end{aligned} \tag{13}$$

It is easy to check that $\nabla_\alpha\Phi$ transforms homogeneously under the group (2) (from there on we consider only infinitesimal parameters):

$$\delta\,(\,\nabla_\alpha\,\Phi\,) \equiv (\,\nabla_\alpha\,\Phi\,)' - \nabla_\alpha\,\Phi = -(\,\nabla_\alpha\,\varrho^\beta\,)\,\nabla_\beta\,\Phi \quad . \tag{14}$$

In the case $n = -\frac{1}{3}$ (minimal supergravity), where H^μ and ϱ^μ are absent, the role of ∇_α is played by Δ_α (13):

$$\delta\,(\,\Delta_\alpha\,\Phi) = -\,(\,\Delta_\alpha\,\lambda^\beta\,)\,\Delta_\beta\,\Phi \quad .$$

Notice the following algebraic properties (most easily proved in the right-handed basis in $\mathbb{R}^{414}$):

$$\{\nabla_\alpha\,,\nabla_\beta\}=0\quad;\quad\{\Delta_\alpha,\Delta_\beta\}=0\quad. \tag{15}$$

Now we come to the definition of our basic building blocks. Consider the quantities

$$\begin{aligned} A &= det\,(\,\frac{1}{4}\,\sigma_a^{\dot\alpha\alpha}\,[\,\Delta_\alpha\,,\overline{\Delta}_{\dot\alpha}\,]\,H^m\,)\ ,\quad A=A^+\quad, \\ B &= det\,(\,\delta_m^n+i\,\partial_m H^n\,)\ ,\quad C=det\,(\,\delta_\alpha^\beta+\Delta_\alpha\,H^\beta\,)\ . \end{aligned} \tag{16}$$

They transform infinitesimally as follows

$$\begin{aligned} \delta A &= (\omega+\overline{\omega}+\gamma)\,A\quad, \\ \delta B &= (\overline{\omega}+l-\gamma)\,B\quad, \\ \delta C &= (\omega+r-R)\,C\ . \end{aligned} \tag{17}$$

Here

$$\begin{aligned} \omega &= \Delta^\alpha\,\lambda_\alpha = det^{-1}\,(\,\Delta_\alpha\,\theta'^\beta\,)-1+O\,(\,\lambda^2\,) \\ \gamma &= \frac{1}{2}\,\partial_m\,(\,\lambda^m+\overline{\lambda}^m\,)-\partial_\alpha\,\lambda^\alpha-\overline{\partial}_{\dot\alpha}\,\overline{\lambda}^{\dot\alpha} \\ l &= \partial_m^L\,\lambda^m-\partial_\mu^L\,\lambda^\mu\ ,\quad r=l^+ \\ L &= \partial_m^L\,\lambda^m-\partial_\mu^L\,\lambda^\mu-\overline{\partial}_{\dot\mu}^L\,\overline{\varrho}^{\dot\mu}\ ,\quad R=L^+\quad. \end{aligned} \tag{18}$$

The Einstein subgroup condition (6) becomes in these terms

$$(3n+1)\,L=2\,n\,l\quad. \tag{19}$$

The Berezinians of changes of variables between different bases in $\mathbb{R}^{414}$ are expressed in terms of the blocks in the following way

$$\begin{aligned} &Ber\,(\,\frac{\partial Z_L}{\partial Z}\,)=B\,\overline{C}^{-1}\quad,\quad Ber\,(\,\frac{\partial Z_R}{\partial Z}\,)=\overline{B}\,C^{-1}\quad, \\ &U=Ber\,(\,\frac{\partial Z_L}{\partial Z_R}\,)=B\,\overline{B}^{-1}C\,\overline{C}^{-1}\quad. \end{aligned} \tag{20}$$

We are going to make heavy use of the blocks defined above for constructing various quantities given their transformation properties.

4 Action formulas

Now it is very easy to find the general action formula for all n (except $n=0$). It is simply the supervolume [18, 13]

$$S=\frac{1}{n\,\kappa^2}\int d^8z\,E \tag{21}$$

and it says that our hypersurface is the minimal one. The density E must compensate the transformations of the volume element:

$$\delta\,E=-\,\gamma\,E\quad. \tag{22}$$

Such a quantity can easily be built from the blocks (17) (taking into account (19))

$$E = A^n \,(\, B\,\overline{B}\,)^{\frac{n+1}{2}}\,(\,C\,\overline{C}\,)^{\frac{3n+1}{2}} \quad . \tag{23}$$

As can be expected, at $n = -\frac{1}{3}$ the blocks C, $\overline{C}$ containing H^μ, $\overline{H}^{\dot\mu}$ disappear.

The formula (23) is valid for $n = 0$ as well but (21) is not an action any more. An indication of this is the dropping out of the block A. The latter is the only one which contains the scalar curvature as a component field. In the case $n = 0$ the supergravity action can be written down in one of the following two ways:

i) $n = 0$: new minimal supergravity. Taking into account the constraint (10), (20) and (23) one finds

$$E = Ber\,(\,\frac{\partial Z_L}{\partial Z}\,) = Ber\,(\,\frac{\partial Z_R}{\partial Z}\,) \quad . \tag{24}$$

Therefore

$$\int d^8 z\, E = \int d^8 Z_L \cdot 1 = 0 \qquad (if\ \ U = 1\ \ ) \tag{25}$$

i.e. the supervolume of $\mathbb{R}^{414}$ vanishes in this case [15, 11], and we cannot use it to write down the action. However, there is an action of the form [11] [1]

$$S^{new\ min} = \frac{1}{\kappa^2} \int d^8 z\, E \ln f \quad . \tag{26}$$

It will be invariant if the quantity f transforms according to

$$\delta f = -\frac{1}{2}\,(l + r)\, f \quad or \quad \delta \ln f = -\frac{1}{2}\,(l + r) \quad . \tag{27}$$

Indeed, taking into accoumt $U = 1$ we have

$$\begin{aligned} \delta \int d^8 z\, E \ln f \;\; &= \;\; -\int d^8 z\, E\, \frac{1}{2}\,(l + r) = \\ &- \;\; \frac{1}{2} \int d^8 Z_L\, l\,(\,\zeta_L\,) - \frac{1}{2} \int d^8 Z_R\, r\,(\,\zeta_R\,) = 0 \quad . \end{aligned}$$

Such a quantity can uniquely be built out of the blocks,

$$f = A^{1/2}\,(\,B\,\overline{B}\,)^{1/4}\,(C\,\overline{C}\,)^{-3/4} \quad . \tag{28}$$

Notice that the important "kinetic" block A is again present in (26).

ii) $n = 0$: the flexible supergravity. As we have explained in sect. 2., in the flexible supergravity we lift the constraint $U = 1$. Then the action (27) ceases to be invariant. The $U(1)$ gauge compensator (11) introduced above is used to restore the invariance. The integral

$$I = \int d^8\, Z_L\, (\ln f + i\,\varphi) \tag{29}$$

is invariant $(\delta\,(\ln f + i\,\varphi) = -l\,(\zeta_L)\,)$ and the action becomes [5]

$$S^{fl.} = \frac{1}{2\kappa^2} \int d^8 z\, E\,[\,U^{1/2}\,(\ln f + i\,\varphi) + U^{-1/2}\,(\ln f - i\,\varphi)\,] \quad . \tag{30}$$

---

[1] It is worthwhile to note that the action (26) can be obtained from Eq. (21) by taking the limit $n \to 0$ [19] and imposing the constraint $U = 1$.

It is important to realize that the compensator φ appears in (30) as a Lagrange multiplier. The variation of φ produces the equation of motion

$$U^{1/2} = U^{-1/2} \tag{31}$$

which coincides with the constraint (10). So, on-shell and in the absence of matter the flexible supergravity is equivalent to the new minimal one. The substantial difference between the new minimal and flexible supergravity becomes clear in the context of matter coupling.

References

[1] K.S.Stelle, P.C.West, Phys.Lett. **74B** (1978) 330.
S.Ferrara, P. van Nienwenhuizen, Phys.Lett. **74B** (1978) 333.

[2] P.Breitenlohner, Nucl.Phys. **B124** (1977) 500.

[3] W.Siegel, J.Gates, Nucl.Phys. **B147** (1979) 77.

[4] V.Akulov, D.Volkov, V.Soroka, Theor.Math.Phys. **31** (1977) 12;
M.F.Sohnius, P.West, Phys.Lett. **105B** (1981) 353.

[5] A.Galperin, V.Ogievetsky, E.Sokatchev, J.Phys. A: Math. Gen. **15** (1982) 3785.

[6] S.Ferrara, L.Girardello, T.Kugo, A. van Frogen, CERN preprint TH 3523 (1983).

[7] C.S.Aulakh, M.Kaku, R.W.Mohapatra, City Univ. of New York preprint (1983).

[8] S.J.Gates, M.T.Grisaru, W.Siegel, Nucl.Phus. **B203** (1982) 189.

[9] E.Witten, Nucl.Phys. **B186** (1981) 513;
B.Zumino, preprint LBL-15819 (1983).

[10] V.Ogievetsky, E.Sokatchev, Phys.Lett. **B79** (1978) 222; Yadernaya Phys. **31** (1980) 264, 821; **32** (1980) 862, 870, 1192;
E.Sokatchev, in "Superspace and Supergravity" eds.
S.Hawking and M.Rocek, Cambr. Univ. Press (1981) 197.

[11] P.S.Howe, K.S.Stelle, P.Townsend, Phys.Lett. **107B** (1981) 420.

[12] K.S.Stelle, preprint LPTENS 82/3 (1982).

[13] S.Deser, B.Zumino, Phys.Rev.Lett. **38** (1977) 1433.

[14] W.Siegel, preprint HUTP-77/AO68 (1977).

[15] B. de Wit, J.W. van Holten, A. van Proeyen, Nucl.Phys. **B167** (1980) 186;
B. de Wit, M.Rocek, Phys.Lett. **109B** (1982) 439;
B.M.Zupnik, Yadernaya Phys. **36** (1982) 779.

[16] B.Zupnik, Talk at Zvenigorod Conference on Group-Theoretic Methods in Physics, Zvenigorod (1982).

[17] P.Howe, K.S.Stelle, P.Townsend, Nucl.Phys. **B214** (1983) 519.

[18] J.Wess, B.Zumino, Phys.Lett. **74B** (1978) 51.

[19] A.A.Rosly, A.S.Schwarz, Yadernaya Phys. **37** (1983) 786.

II. Differential Geometric Methods

ON THE USE OF ASSOCIATIVE ALGEBRAS IN DIFFERENTIAL GEOMETRY

Armin Uhlmann

Karl-Marx-University
Dep. of Physics and NTZ
7010 Leipzig, Am Karl-Marx-Platz
German Democratic Republic

ABSTRACT

We like to explain why and how to use certain associative algebras in the differential geometry of smooth manifolds.

1. Introduction

Let me start with some short remarks concerning the use of algebras in topology. More than 40 years ago I. Gelfand [1] has shown how every unital C^*-algebra uniquely determines (up to topological equivalence) a compact Hausdorff space by its maximal ideals, thus giving a new input to the von Neumann - Murray theory of operator algebras. Moreover, regarding **C(T)**, the algebra of all complex-valued continuous functions defined on a compact Hausdorff space, **T**, as an abstract algebra, one can recover its "natural" C^*-structure. Hence by Gelfand's theo= rem, every topological property of the space **T** is uniquely reflected in structural properties of the algebra **C(T)**. One may, therefore, if one like to do so, establish a sort of "dic= tionary" in which to every topological property of **T** there is written down one or several ways of "translating" it into expres= sions only using knowledge of the abstract algebra **C(T)**.
Of course, one could go equally well the other way round: To look at algebraic properties of **C(T)**, and ask for the topolo= gical meaning of them within **T**.

However, buy an ordinary dictionary and try to translate something from the German into the Russian language. It doesn't work that easy - and the same is with our hypothetical mathe= matical "dictionary". A one-to-one translation of a topological property is not obliged to fit well into the language of the al= gebra, and often one has to modify the concepts in question.

Thus the space **T** and the algebra **C(T)** are really different aspects of the same "thing". To look at both may be a source of intuition. Furthermore, having "translated" topolo= gical concepts into the language of the algebra **C(T)**, one won= ders wether these "translations" should not work in the case of some non-commutative algebras too. This, indeed, is one way in approaching the question what "non-commutative geometry" is or should be. And indeed, one may hope to find along those pathes a new one that reaches again Quantum Physics, i.e. the very source of operator algebras.

Let us now ask how to deal with differentiability by means of algebras. The quite important point is: One has to leave the class of C^*-algebras.
Suppose $\mathbb{M}$ is a differentiable manifold (countable at infinity). The set $\mathbf{C}^{(k)}(\mathbb{M})$ of all complex-valued k-times differentiable functions is not carrying a C^*-topology for k biger than 0. If k is finite one can make these algebras Banach algebras. But if k is infinite, i.e. for smooth functions, this is impos= sible. On the contrary, $\mathbf{C}^{(\infty)}(\mathbb{M})$ carries naturally a nuclear topology. One may construct this topology as following: Let D be a smooth differential operator and $\mathbb{K}$ a compact subset of $\mathbb{M}$. Then

$$\| f \|_{\mathbb{K},D} := \sup_{p \in \mathbb{K}} | (Df)(p) |$$

is a seminorm in $\mathbf{C}^{(\infty)}(\mathbb{M})$. The collection of seminorms obtai= nable in this way defines the topology. Though we have intro= duced this topology by the aid of $\mathbb{M}$, it is possible to do so "abstractly" by using the algebra $\mathbf{C}^{(\infty)}(\mathbb{M})$ as an abstract ob= ject. That this is so comes from the following fact: The algebra of smooth differential operators is generated over the smooth functions by the derivations (see later).

In the following we shall give some elementary proper= ties of the algebra $\mathbf{C}^{(\infty)}(\mathbb{M})$ and their geometrical interpreta= tion. Similar considerations can be done with algebras the ele= ments of which are smooth functions with values in matrix- or Grassmann algebras. Further, as an more general object, one may consider the algebra of smooth sections of a differentiable fibre bundle the fibres of which are finite-dimensional associa= tive algebras. It should be clear that only a first indication how things work can be given, and for the experts of the domain in questions this is, more or less, well known.

2. Points.

The set $\mathbb{I}_p$ of all those elements of $\mathbf{C}^{(\infty)}(\mathbb{M})$ vanishing at a given point $p \in \mathbb{M}$ is a maximal ideal of this algebra.

2.1. Proposition. Let $\mathbb{I}$ be a maximal ideal of $\mathbf{C}^{(\infty)}(\mathbb{M})$. The following conditions are mutually equivalent:

(a) There is a point $p \in \mathbb{M}$ with $\mathbb{I} = \mathbb{I}_p$

(b) $\mathbb{I}$ is closed in the natural topology.

(c) $\mathbb{I}$ is finitely generated.

(d) It is $\mathbf{C}^{(\infty)}(\mathbb{M}) / \mathbb{I}$ isomorphic to $\mathbf{C}$ (complex numbers).

(e) There is a state u with $u(\mathbb{I}) = 0$.

For the algebra at hand a state is given by a probability mea= sure with compact support on $\mathbb{M}$, and $u(.)$ denotes the ex= pectation value.
We shall not give proofs of this (rather simple, indeed) and the following assertions. The reader can find a rich and essential selection of facts and proofs in the book of Malgrange [2]. The proofs of most of the assertions of this section can be found there (or at least their crucial idea).

Let us call every maximal ideal satisfying one (and hence all) of the conditions of proposition 2.1 a "point ideal" or simply a "point" of $\mathbf{C}^{(\infty)}(\mathbb{M})$.

2.2. Remark. Let $\mathbb{I}$ be any maximal ideal. At first restricting it to the algebra of bounded smooth functions, and then exten= ding it to the C^*-algebra of bounded continuous functions on $\mathbb{M}$ we see by Gelfand's theorem: The maximal ideals of $\mathbf{C}^{(\infty)}(\mathbb{M})$ correspond one-to-one to the points of the Čech-compactification of $\mathbb{M}$. Hence a maximal ideal of the algebra of smooth functions is either a point ideal or it characterizes a boundary point of the Čech-compactification.
Further, the intersection of all maximal ideals which ar not point ideals consists of all smooth functons with compact sup= port.
Thus $\mathbb{M}$ is compact iff every maximal ideal is a point ideal.

2.3. Proposition.

(i) Two manifolds $\mathbb{M}_1$ and $\mathbb{M}_2$ are diffeomorphic if and only if $\mathbf{C}^{(\infty)}(\mathbb{M}_1)$ and $\mathbf{C}^{(\infty)}(\mathbb{M}_2)$ are algebraically isomorph.

(ii) Let $\mathbb{N}$ be a submanifold of $\mathbb{M}$ and let $\mathbb{I}_{\mathbb{N}}$ denote the ideal of all smooth functions defined on $\mathbb{M}$ and vanishing on $\mathbb{N}$. There are natural homomorphisms

$$0 \to \mathbb{I}_{\mathbb{N}} \longrightarrow \mathbf{C}^{(\infty)}(\mathbb{M}) \longrightarrow \mathbf{C}^{(\infty)}(\mathbb{N}) \longrightarrow 0$$

This sequence is an exact one iff $\mathbb{N} \cap \mathbb{K}$ is compact for every compact subset $\mathbb{K}$ of $\mathbb{M}$.

It is worthwhile to notice in this connection the possibility of de= fining "smooth manifolds with singularities" by performing factor-algebras of $\mathbf{C}^{(\infty)}(\mathbb{M})$ with "suitable" ideals.
The next assertion is a basic one in reflecting the differentiable structur of $\mathbb{M}$ by its algebra of smooth functions.

2.4. Proposition. Let $\mathbb{I} = \mathbb{I}_p$, $p \in \mathbb{M}$, a point ideal.
It is $f \in \mathbb{I}^{n+1}$ if and only if f and all its partial derivatives up to order n vanish at p.

There are some immediate consequences.

For every point ideal $\mathbb{I}_p$ the factor algebra

$$\mathbb{C}^{(\infty)}(\mathbb{M}) \;/\; \mathbb{I}_p^{(m+1)}$$

is the linear space of m-jets attached at $p \in \mathbb{M}$. Hence

$$\mathbb{I}_p \;/\; \mathbb{I}_p^{2}$$

is the fibre of the cotangent bundle at $p \in \mathbb{M}$.
The fibre of the tangent-space at p is, therefore, the dual, i.e. the set of linear forms, of that factor algebra. This fibre can be identified with the set of all such linear forms on $\mathbb{C}^{(\infty)}(\mathbb{M})$ which satisfy

$$u(\underline{1}) = 0 \quad \text{and} \quad u(\mathbb{I}_p^{2}) = 0 \quad .$$

A (smooth) vector field, X , is now a (smooth) section

$$p \longrightarrow X_p \in (\mathbb{I}_p / \mathbb{I}_p^{2})^{*}.$$

All these concepts have to be understood in their complex version for we work with algebras over the complex numbers, $\mathbb{C}$.

Let $\mathbb{J}$ be an arbitrary but closed in the natural topology ideal. It is not difficult to see that by obvious modification one can introduce the concept of point, of m-jet, of co-tangent bundel,... within the algebra $\mathbb{C}^{(\infty)}(\mathbb{M}) / \mathbb{J}$.
More generally, one can define for unital C^*-algebras $\mathbb{A}$ the concept of point ideals by selecting those maximal ideals of $\mathbb{A}$ which are of finite co-dimension. By the aid of the powers of these point ideals the concept of m-jet, cotangent-bundle, and so on can algebraically be given. Of course, not for all such algebras this gives very meaningful concepts. But it does so for example in algebras of smooth matrix- or Grassmann-valued functions on smooth manifolds, and in many other geometrically "meaningful" algebras.

Now let us return to the algebra $\mathbb{C}^{(\infty)}(\mathbb{M})$. Let us define

$$\mathbb{I}_p^{(\infty)} = \bigcap_n \mathbb{I}_p^{n} \; .$$

This ideal consists of all such functions the derivatives of all orders of which vanish at p . Choosing a local coordinate system, the factor class $g + \mathbb{I}_p^{(\infty)}$ corresponds to the formal power sery of g at p with respect to this coordinate system. One knows that

$$\mathbb{C}^{(\infty)}(\mathbb{M}) \;/\; \mathbb{I}_p^{(\infty)}$$

is isomorphic to the algebra of formal power series in dim $\mathbb{M}$ variables.

Consider now an ideal $\mathbb{J}$. For every $p \in \mathbb{M}$ it generates an ideal in the algebra of formal power series attached to p in the manner described above by the the homomorphism

$$\mathbb{J} \longrightarrow \mathbb{J} + \mathbb{I}_p^{(\infty)} \;/\; \mathbb{I}_p^{(\infty)} \; .$$

It is one of the astonishing discoveries of Whitney [3] asserting that those induced ideals in the algebras of power series already completely characterize J if (and only if) J is closed in the natural topology of the algebra of smooth functions.

2.5. Theorem (Whitney)
Let J_1 and J_2 be two closed ideals of $C^{(\infty)}(M)$. Then

$$J_1 = J_2 \quad \text{iff} \ \forall\, p \in M: \quad J_1 + I_p^{(\infty)} = J_2 + I_p^{(\infty)} .$$

There is a further remarkable ideal attached to a given point $p \in M$: Let

$$I_p^{(min)}$$

denote the ideal consisting of all such smooth functions f which vanish in some neighbourhood of p. This ideal is minimally primary and is defined algebraically as following.

2.6. Proposition: Let $I' \subseteq I_p$. Assume that I' is not contained in any maximal ideal different from I_p. Then

$$I_p^{(min)} \subseteq I' .$$

The factor algebra

$$C^{(\infty)}(M) \ / \ I_p^{(min)}$$

is the algebra of the <u>germs of differentiable functions</u> at p.

As a further "translation" in the sense of the introduction we shall indicate how to handle folitations.

Let F be a subalgebra of $C^{(\infty)}(M)$. This algebra is called <u>foliating</u> if

a) it is a unital *-algebra, i.e. it contains the identity $\underline{1}$ of the algebra of smooth functions, and with every function its complex conjugate.
b) if $f \in F$ and f^{-1} exists in $C^{(\infty)}(M)$ then $f^{-1} \in F$.

Let us assume F is foliating. Then to every point $I°$ of F there are points p M with $I° \subseteq I_p$. The set of all these points perform a <u>leaf</u>, the leaf attached to $I°$. The foliating condition guaranties: Every point of M (and hence every point of the algebra) belongs to just one leaf, and the set of all leafs is parametrized by the point ideals of F. The parametrization can be called smooth if F is algebraically isomorphic to an algebra of smooth functions over a certain smooth manifold.
Two foliating algebras may give the same foliation. However, given a foliation F, the subset of all functions which are constant along each leaf is a foliating algebra containing F and determining the same foliation. Hence if there is any there is also a maximal algebra defining a given foliation.
These notations extend to more general algebras.

3. Derivations et cetera.

Let $\mathbb{A}$ be an algebra over the complex numbers, with unit element, and not necessarily commutative.
A linear map

$$L \; : \; \mathbb{A} \longrightarrow \mathbb{A}$$

is called a <u>derivation</u> of $\mathbb{A}$ if for all $a,b \in \mathbb{A}$ it is

$$L(ab) = (La)\, b + a\, Lb .$$

Such a definiton would be rather catastrophical for C^*-algebras. With these algebras one cannot require the domain of definition of $L_{(\infty)}$ to be the whole algebra. On the contrary, for the algeb= ra $\mathbf{C}^{(\infty)}(\mathbb{M})$ and some other ones of nuclear type this is quite natural.
The derivations of $\mathbb{A}$ form a comlex Lie algebra: with L_1, L_2 also $L_1L_2 - L_2L_1$ is a derivation. Let us call the Lie algebra of all derivations of $\mathbb{A}$ by

$$\text{Deriv } \mathbb{A} .$$

We now consider a representation

$$\phi \; : \; a \longrightarrow \phi(a) \;, \quad a \in \mathbb{A}$$

with the linear representation space $\mathbb{L}$. We want not to consi= der the topological properties of representations here. Thus we consider ϕ to be a homorphism of the algebra into the algebra End $\mathbb{L}$ of all linear maps from $\mathbb{L}$ into $\mathbb{L}$. The question is now how to "transport" the derivation to the representation space. To this end one introduces the concept of <u>co-derivation.</u>

Let ϕ be a representation. A co-derivation μ is a linear map

$$\mu \; : \; \text{Deriv } \mathbb{A} \longrightarrow \text{End } \mathbb{L}$$

satisfying the following relation for all $a \in \mathbb{A}$ and $x \in \mathbb{L}$

$$L^{\mu}\, \phi(a)\, x = \phi(\,La)\, x + \phi(a)\, L^{\mu}\, x .$$

Here the result of an application of the linear map μ on a derivation L is denoted by L^{μ}. It is not always wise to require the full Lie algebra Deriv $\mathbb{A}$ as the domain of defini= tion of μ. But for simplicity we shall do so here.
One should emphasize the dependence of μ from a given ϕ, for without specifying a representation the notation of co-deri= vatives is logically incomplete. In the most important applica= tions there is a natural action of $\mathbb{A}$ on the representation space. A good example is the action of $\mathbf{C}^{(\infty)}(\mathbb{M})$ on the sec= tions of a complex vector bundle. Thus if there is no danger of confusion we shall later on abbreviate

$$\phi(a)\, x \quad \text{by} \quad a\, x \quad (\text{ or } \; ax\,) .$$

Let now $\mathbb{J}$ be an ideal of $\mathbb{A}$. We abbreviate the set of all $\phi(b)\, x$ with $b \in \mathbb{J}$ and $x \in \mathbb{L}$ by $\mathbb{J}\,\mathbb{L}$. It is easy to see that for $n = 0,1,2,\ldots$ one has

$$L^{\mu} : \mathbf{J}^{n+1} \mathbb{L} \subseteq \mathbf{J}^{n} \mathbb{L} \quad , \text{ with } \mathbf{J}^{o} = \mathbb{A} .$$

The co-derivation μ is said to be "<u>of order k with respect of **J**</u>" if

$$L\,\mathbf{A} \subseteq \mathbf{J}^{k+1} \quad \text{implies} \quad L^{\mu}\,\mathbb{L} \subseteq \mathbf{J}\,\mathbb{L}$$

and if k is the smallest integer with this property.
Remember now that a point of $\mathbf{A}$ is a maximal ideal with finite codimension.
We shall say from a co-derivation μ it is <u>k-local</u> if μ is of order k with respect to every point of $\mathbf{A}$. In the case $k = 0$ we simply speak of a <u>local co-derivation</u>.
Local co-derivations are of special importance for they are the equivalent to the affine connections.

Though a co-derivation is defined via a representation of the algebra by linear operators in a linear space, it is by no means true that the co-derivation is an homomorphism of Lie algebras, generally. The deviation of μ from being a Lie homomorphism from Deriv $\mathbf{A}$ into End $\mathbb{L}$, where End $\mathbb{L}$ is considered as a Lie algebra with the commutator braket as the Lie multiplication, will be "estimated" by its curvature.
The <u>curvature</u> of a co-derivation μ is an antisymmetric and bilinear form on Deriv $\mathbf{A}$ with values in End $\mathbb{L}$. It is defined by the relation

$$[L_1{}^{\mu}, L_2{}^{\mu}] = [L_1, L_2]^{\mu} + \Omega(L_1, L_2) .$$

From this it is clear that μ is an homomorphism of Lie algeb= ras iff $\Omega = 0$.

Now we have explained a (part of a) rather abstract scheme that is attached to algebras and their representation. An algebra for which these concepts are meaningful is our old $\mathbf{C}^{(\infty)}(\mathbb{M})$. Let us now go once more to these concepts above, and let us look how we can identify them with geometrical ones.

If X is a vector field of $\mathbb{M}$, then we can construct the Lie derivative L_X. L_X operates on smooth functions as as a derivative.
The point is now: There are no other derivatives of $\mathbf{C}^{(\infty)}(\mathbb{M})$ as those coming from vector fields. The set of all derivatives of $\mathbf{C}^{(\infty)}(\mathbb{M})$ is identical with the set of all Lie derivatives L_X formed by complex vector fields X of $\mathbb{M}$. Indeed, let L be a derivation of our algebra and $p \in \mathbb{M}$. Then X_p is identified with the following linear form over the cotangent fibre

$$X_p : f - f(p)\underline{1} + \mathbf{I}_p{}^2 \longrightarrow (Lf)(p) .$$

Let us remain a moment with derivations. If X is a real vector field it generates locally a group germ of a one-parameter diffeomorphism group. Its orbits give a foliation of $\mathbb{M}$. Just this foliation can be given by a foliating subalgebra. This subalgebra consists of all f with $Lf = 0$.

More generally, given m real vector fields, one can construct with the coresponding Lie derivatives $L_1, \ldots, L_m$ a foliating subalgebra of $C^{(\infty)}(M)$ by the conditions

$$L_1 f = \ldots = L_m f = 0 .$$

At those points of M at which the sytem of derivations is or generates by its Lie closure an integrable system of partial differential equations, the leafs of the foliation coincides with the integral manifolds.

Let us now consider some co-derivations of $C^{(\infty)}(M)$. At first we have to specify the representation. The most simple one is to consider the algebra itself as the representation space on which the algebra acts by (left) multiplication:

$$\phi \; : \quad \phi(a)\, b = ab \quad , \qquad a, b \in C^{(\infty)}(M) .$$

Let μ denote a co-derivation with respect to this representation. Then

$$L^{\mu}\, b = L^{\mu}(b\, \underline{1}) = L\, b + b\, L^{\mu}\, \underline{1}$$

by the general rules for co-derivations. In this case, the physicist would say "in the case of scalar fields", the co-derivation is characterized by a linear map

$$L \longrightarrow L^{\mu}\, \underline{1} \quad , \quad L \in \text{Deriv}\; C^{(\infty)}(M)$$

into the linear space of smooth functions on M.
Let us now assume μ to be local. If then a vector field vanishes at p the co-derivative has to vanish at the same point. That means $(L^{\mu}\, \underline{1})(p)$ depends only on the vector attached at p, and can, therefore, be given by a covector at p. Hence there is a covector field A such that

$$L_X^{\mu}\, b = L_X\, b + (X^k A_k)\, b$$

where local expressions for the tangent and cotangent fields X and A has been used. It is now plain to see

$$(X,Y)\, b = X^k Y^j (A_{k,j} - A_{j,k})\, b$$

i.e. the curvature is multiplication "with the electromagnetic field strength given by the potential A" as physicists would perheps say. (We use comma notation for partial derivatives.)

But what if μ is not local but n-local ? Then the local expression for a coderivative can be expressed by several "generalized potentials" A as follows:

$$L_X^{\mu} = L_X + A_j X^j + A_j^{\,k} X^j_{\;,k} + A_j^{\,k_1 \ldots k_n} X^j_{\;,k_1 \ldots k_n}$$

i.e. the Lie derivative is complemented by a rather complicated form of the function $L^{\mu}\, \underline{1}$ that acts as a multiplication operator. The curvature is a complicated sum

$$\Omega(X,Y) = \Omega_0(X,Y) + \Omega_1(X,Y) + \ldots + \Omega_n(X,Y)$$

where

$$\Omega_i = A^{k_1 \ldots k_i}{}_{,k} \left(X^k Y^j{}_{,k_1 \ldots k_i} - Y^k X^j{}_{,k_1 \ldots k_i} \right) +$$

$$+ A^{k_1 \ldots k_i}{}_{,j} \Big(X^k (Y^j{}_{,k})_{,k_1 \ldots k_i} - Y^k (X^j{}_{,k})_{,k_1 \ldots k_i} +$$

$$+ (X^k (Y^j{}_{,k}))_{,k_1 \ldots k_i} - (Y^k (X^j{}_{,k}))_{,k_1 \ldots k_i} \Big).$$

This, again, is a local expression valid with respect to some coordinate system, and for the partial derivatives with respect to this coordinate system the comma notation is used. Clearly, a k-local co-derivation with k different from zero is a rather complicated object.

Now let $\mathbf{L}$ be the linear space of the smooth sections of a smooth vector bundle with base space $\mathbf{M}$. Canonically, $\mathbf{L}$ is a representation space of $\mathbf{C}^{(\infty)}(\mathbf{M})$ by (left) multiplication.

Every affine connection defines in an obvious manner a certain co-derivation. This construction exhausts just <u>all local co-derivations</u> in the case at hand.
To write down the form of n-local co-derivations is a cumbersome but rather straightforward task.

Now I like to add some more general remarks.

The relation between vector fields and derivatives does not remain valid for more general algebras $\mathbf{A}$. A derivation induces a map

$$L : \mathbf{I} / \mathbf{I}^2 \longrightarrow \mathbf{A} / \mathbf{I}.$$

If $\mathbf{I}$ is a point but $\mathbf{A} / \mathbf{I}$ is not isomorphic to $\mathbf{C}$ this map does not give an element of the dual of the cotangent fibre $\mathbf{I} / \mathbf{I}^2$. By the very definition of "point" the most general situation is for $\mathbf{A} / \mathbf{I}$ to be a full matrix algebra.

As an example we shortly consider the case
$\mathbf{A}$ = algebra of smooth matrix-valued function of order n defined on a smooth manifold $\mathbf{M}$.
A derivation L then is of the form

$$a \longrightarrow L\, a \quad = \quad L_X\, a \; + \; ba - ab$$

where the Lie derivative acts on the entries of the matrix. The derivative is called <u>inner</u> iff $X = 0$. The inner derivatives form a normal sub-Lie algebra of Deriv $\mathbf{A}$.
The co-derivations $\boldsymbol{\mu}$ associated with left multiplication within the algebra ("scalar" case a la Higgs) are defined by

$$\boldsymbol{\mu} : \quad L \longrightarrow L^{\boldsymbol{\mu}}\, \underline{1}.$$

It is $L^{\mu}\underline{1} = 0$ for all inner derivations L. It is for this reason that we get for <u>local co-derivations</u> the correct transformation properties of connections for gauge theories in the expression

$$L^{\mu}\underline{1} = X^i A_i \ , \quad L = L_X + L_{inner}$$

The local co-derivations of **A** are, therefore, in a one-to-one correspondence to the gauge potentials of gauge theories, and the curvatures correspond as usual to the field strength of the potentials. To obtain "all" gauge fields one has to define the co-derivations on a suitable Lie subalgebra of Deriv **A** which is in physical applications often much smaller than Deriv **A** .

A further remark concerns the so-called <u>modified derivations</u>. Let w be a linear map of some algebra **A** into itself. (Here we have <u>not</u> in mind the algebra of smooth matrix functions!) A <u>w-derivation</u> is a linear map of **A** into **A** with

$$L\,(ab) = (La)\,b + (wa)\,Lb$$

Under very weak assumption (L **A** should contain at least one element that is not a divisor of the zero) one can then conclude that w has to be an automorphism of **A** .

Of particular importance is the case where w is a distinguished reflection, i.e. w w = identity , in which circumstances one now calls w a <u>superstructure</u>. If w is a superstructure of **A** then one considers those w-derivations for which

$$L\,w + w\,L = 0$$

is valid. Together with the ordinary derivations they form a so-called super Lie algebra (a Z_2-graded Lie algebra).

It is now plain to define <u>co-w-derivations</u> and their curvature forms. A good canditate to study the situation is the algebra of smooth functions on a manifold with values in a Grassmann algebra. Due to the appearance of a non-trivial radical the notation of "k-locality" of a co-derivation has to be used here with some caution in order not to exclude interesting examples.

References

1) Gelfand, I.M., Mat.Sb. 9 (1941), 3-24
2) Malgrange, B.: Ideals of Differentiable Functions. Oxford University Press, 1966.
3) Withney, H., Am.J.Math. 70 (1948), 635-658

Spinors and Moduli of Einstein Metrics on Kähler Simply Connected Manifolds with a Canonical Class $K \equiv 0$

Andrei Nikolov Todorov
Sofia 1113,
Institute of Mathematics,
ul."Acad. G. Bonchev" 8, P.B. 373
Bulgaria

1 Introduction

In his article [1] Marcel Berger wrote: "Il y a deux questions principales qui contiennent toutes les autres: Q1. Quelles sont les variétés compactes qui admettent au moins une mètrique d'Einstein? Q2. Sur une telle variété, decrire l'ensembles des mètriques d'Einstein?"

In this article we will deal with the second question. More precisely the following theorems are announced: **Theorem I.** Suppose that X is simply Kähler connected manifold with canonical class $K \equiv$ and $H^0(X, \Omega^2) = 0$, then the space of all non isomorphic Einstein metrics is isomorphic to the moduli space of all non-isomorphic complex structures on X. Both spaces have complex dimension equal to dim $H^1(X, \Omega^{n-1})$.

Theorem II. Suppose that X is a simply connected symplectic manifold, then the moduli space of all non-isomorphic Einstein metrics is an open and everywhere dense subset in

$$\Omega = SO(3, b_2 - 3)/SO(3) \times SO(b_2 - 3).$$

Yau's solution of Calabi's conjecture shows that a Kähler-Einstein

metric on Kähler manifold with canonical class $K \equiv 0$ always exists.

The proof of theorem I is relatively easy, while the proof of theorem II is based on much deaper theorems and is not trivial. Here we give only the ideas of the proof.

2 The structure Theorem

Let me first set up some terminology. A manifold is always assumed to be connected. By a Kähler manifold we mean a complex manifold which admits at least one Kähler metric.

The structure theorem for manifolds with $K \equiv 0$ goes back to Calabi [3]. A stronger form was proved by Bogomolov [4]. Finally the proof by Yau of the Calabi conjecture made possible to give an easy proof of the strongest possible statement. See the article by M.L.Michelsohn [5].

THEOREM. Let X be a compact manifold with $K \equiv 0$ I) The universal covering of X is isomorphic to a product $\mathbb{C}^K \times \prod_i V_i \times \prod X_j$ where

a) V_i is a simply connected projective manifold, of dim ≥ 3, with trivial canonical bundle, such that $H^0(V_i, \Omega_i^p) = 0$ for $0 < p < dimV_i$.

b) X_j is a simply connected manifold (Kähler), admitting a holomorphic two-form φ_j which is everywhere non-degenerate. Any holomorphic form on X_j is (up to a scalar) a power of φ_j.

This descomposition is unique, up to the order of V_i's and of X_j's.

2) there exists a finite étale cover $\tilde{X}$ of X which is isomorphic to a product $T \times \prod V_i \times \prod Xj$, where T is a complex Torus.

Let us give a sketch of the proof, referring to [5] for details. According to Yau's theorem, X carries a Ricci flat metric. Kähler of course. The de Rham theorem [6] implies that the universal covering of X is isomorphic as a Kähler manifold to a product $C^K \times \prod_i M_i$, where for each i the manifold M_i has irreducible holonomy. Moreover M_i is compact by the Cheeger Gromoll theorem [7]. Since M_i is Ricci flat, its holonomy group H_i is contained in SU(m_i). The list of holonomy groups given by Berger in [8] leaves only two possibilities for H_i, namely $H_i \simeq SU(m_i)$ and $H_i \simeq Sp(m_i/2)$ (if m_i is even).

We now consider holomorphic forms on M_i. The Bochner principle (every holomorphic form on a Ricci flat manifold is parallel with the respect to the connection of the Ricci-flat metric) implies that the holomorphic p-forms on M_i is isomorphic to the space of those p-forms at a given point which are invariant under H_i. From the representation theory of the unitary and symplectic groups one deduce easily that M_i satisfies property a) if $H_i \cong SU(m_i)$ and property b) if $H_i = Sp(m_i/2)$. This proves the existence of the decomposition of I) The unicity is deduced easily from the de Rham decomposition and the unicity of the Ricci flat metric in a given cohomology class. Finally 2) follows from the classical Bieberbach theorem.

For obvious reasons, manifolds satisfying property a) will be called special unitary, while those satisfying b) will be called symplectic

3 Harmonic Spinors and moduli of Einstein metrics on special unitary and Symplectic manifolds

Let $CL(X)$ denote the Clifford bundle of X. This is a bundle over X whose fibre at a point x of X is the Clifford algebra $CL(T_x(X))$ of the tangent space at x. There is a canonical embedding $T(X) \subset CL(X)$. Furthermore, the riemannian metric and connection extend to $CL(X)$ with the properties:

$$\nabla(\varphi \cdot \psi) = (\nabla\varphi) \cdot \psi + \varphi \cdot (\nabla\psi) \tag{3.1}$$

for all sections $\varphi, \psi \in \Gamma(CL(X))$.

Suppose that $S \longrightarrow X$ is a bundle of left modules over the bundle of algebras $CL(X)$. We assume, furthermore, that S is furnished with the metric and an orthogonal connection ∇ such that:

$$\text{the module multiplication } e: S_x \longrightarrow S_x \tag{3.2}$$

by a unit vector $e \in T_x(X)$ is an isometry $\forall\, x \in X$.

$$\nabla(\varphi \cdot \sigma) = (\nabla\varphi) \cdot \sigma + \varphi \cdot (\nabla\sigma) \;\; \forall \varphi \in \Gamma(CL(X)),\ \sigma \in \Gamma(S) \tag{3.3}$$

Under these assumptions we define the (generalized) Dirac operator $D : \Gamma(S) \longrightarrow \Gamma(S)$

$$D = \sum_{k=1}^{n} e_k \cdot \nabla_{e_k} \tag{3.4}$$

where $e_1, e_2, \cdots, e_n$ denotes any orthonormal basis of the space $T_x(X)$ at each point x. It is not difficult to see that D is an elliptic operator.

Suppose that X is a spin manifold of dimension $2n$ and let $S = \mathcal{S}$ be the complex bundle of spinors over X with its canonical riemannian connection, i.e. let $P_{SO}(X)$ denote the bundle of oriented orthonormal frames on X. Let $P_{Spin}(X)$ be a principal Spin bundle over X with a given Spin -equivariant covering map $\xi \ : \ P_{Spin}(X) \longrightarrow P_{SO}(X)$. ξ is called a spin -structure on X. Consider $Spin(2n) \subset Cl(\mathbb{R}^{2n})$ and recall that $Cl(\mathbb{R}^{2n}) \otimes \mathbb{C} \cong Hom(\mathbb{C}^{2^n})$, so we get a representation

$$\Delta \ : \ Spin(2n) \longrightarrow Hom(\mathbb{C}^{2^n}).$$

$\mathcal{S}$ is defined to be the associated vector bundle:

$$\mathcal{S} = P_{Spin}(X) x_{\Delta} \mathbb{C}.$$

The Dirac operators are self-adjoint and so their index must be zero. If dim $X = 2m$ then $D : \Gamma(\mathcal{S}) \longrightarrow \Gamma(\mathcal{S})$canonically gives rise to a restricted operator D^+ whose index in the compact case will be in general non-zero. The construction of D^+ is as follows: Consider the volume form $\omega \in Cl(X)$ defined by $\omega = i^m e_1 e_2 \cdots e_{2m}$, where $e_1, e_2, \cdots, e_{2m}$ is a local tangent frame field. It is easy to see that $\omega^2 = 1$, $\nabla\omega = 0$ and $\omega e = -e\omega$ for any $e \in T^*(X)$. From these facts it follows that there is a parallel orthogonal splitting: $\mathcal{S} = \mathcal{S}^+ + \mathcal{S}^-$ into the $+1$ and -1 eigenbundles for the left multiplication by ω. Hence the restriction of D gives a pair of elliptic operators

$$D^+ \ : \Gamma(\mathcal{S}^+) \longrightarrow \Gamma(\mathcal{S}^-) \text{ and } D^- \ : \Gamma(\mathcal{S}^-) \longrightarrow \Gamma(\mathcal{S}^+).$$

Let $H^+ = kerD^+$ and $H^- = kerD^-$, then from the ATIYAH-SINGER Theorem we get:

$$Index(D^+) = dimH^+ - dimH^- = \hat{A}(X).$$

Next we will compute $\hat{A}(X)$ for a) X simply connected special unitary, and b) X is a simply connected symplectic manifold. Namely we will prove the following theorem:

Theorem 3.1. Suppose that X is a simply connected manifold. If: a) X is a special unitary complex manifold then $\hat{A}(X) = 2$ if dim$X = 2n$. b) X is a symplectic manifold then $\hat{A}(X) = n$.

Proof: N.Hitchin proved the following theorem in [9]: Let X be a compact Kähler manifold, then

1) X is Spin iff the canonical bundle K has a square root, i.e., $K \cong L^2$.

2) There is a one-to-one map between spin structures and the holomorphic square roots of K.

3) Under this correspondence $H^+ \cong H^{evev}(X, O_X(L))$ and $H^- \cong H^{odd}(X, O_X(L))$.

In both cases a) and b) since $K \cong O_X, L \cong O_X$, so from the theorem stated above it follows that dim $H^+ = \sum \text{dim}H^0(X, \Omega^{2p})$ and dim $H^- = \sum \text{dim}H^0(X, \Omega^{2p+1})$. Theorem 3.1 now follows from the structure theorem.

Q.E.D

From the Lichnerovich's formula:

$$D^2 = \nabla^*\nabla + \frac{1}{4}r, \text{ where}$$

$$\nabla^*\nabla = -\sum(\nabla_{e_k}\nabla_{e_k} - \nabla_{\nabla_{e_k}e_k}),$$

and r is the scalar curvature, it follows that on a simply connected manifold which is either special unitary of even complex dimension or symplectic one, there are harmonic spinors that are parallel to an Einstein metric. Using representation theory it is not difficult to prove that if on an even dimensional real manifold there exists a harmonic parallel spinor then the holonomy group of this manifold can be reduced tu $SU(n)$. See [9]. From this fact one can deduce that any Einstein metric on a simply connected special unitary or symplectic manifold is a Kähler-Einstein metric with respect to some complex structure. So we can state the following theorem :

Theorem 3.2. If X is an even dimensional, simply connected manifold which is special unitary, then the space of all non-isometric Einstein metrics is isomorphic to the moduli space of all non-isomorphic complex structures on X. Both spaces have dimension equal to dim $H^1(X, \Omega^{n-1})$.

Theorem 3.3. Suppose that X is a simply connected symplectic manifold, then the moduli space af all non -isometric Einstein metrics is an open and everywhere dense subset in

$$\Omega = \Gamma \setminus SO(3, b_2 - 3)/SO(3) \times SO(b_2 - 3)$$

where Γ is a discrete group acting on $SO(3, b_2-3)/SO(3) \times SO(b_2-3)$, and it is defined in [10] as a sub-group in $Aut(H_2(X, Z))$, which preserve a scalar product of type $(3, b_2 - 3)$.

The proof of 3.3 is based on the proof of the global Torelli problem for symplectic manifolds, proved in [11] and the uniqueness of the Kähler-Einstein metric in a fixed class of cohomology.

Notice that $K3$ surfaces are special symplectic manifolds. In case of $K3$ surfaces we know much more precise theorems of type 3.3. See [12]. In Theorem 3.3 we do not have a nice description of the complement of the points in Ω to the moduli space of all non-isometric Einstein metrics, while in the case of $K3$ surfaces we have a complete picture.

References

[1] M.Berger. Rapport sur les variétés d'Einstein, Asterisque **80** pp. 5 - 19 (1980).

[2] S.T.Yau. Comm. Pure and Appl. Math., **31**, 339 - 411 (1978).

[3] E.Calabi. On Kähler manifolds with vanishing canonical class. Algebraic geometry and topology, a symposium in honor of S.Lefshetz, Princeton Univ. Press 78 - 89 (1955).

[4] F.Bogomolov. On the decomposition of Kähler manifolds with trivial canonical class., Math. Sbor. USSR **22** 580 - 583 (1974).

[5] Michelson. Clifford and Spinor cohomology on Kähler manifolds. Amer. J. of Math. **102** 1083 - 1146 (1980).

[6] Kobayashi - Nomizu. Foundations of Differential Geometry I and II, J Wiley, N.Y. (1969).

[7] J.Cheeger, D.Gromoll. The splitting theorem for manifolds of non-negative Ricci curvature. Journal of Diff. Geometry **6** 119 - 128 (1971).

[8] M.Berger. Bull. Soc. Math. France **83** 235 - 274 (1955).

[9] N. Hitchin. Harmonic Spinors. Adv. in Math. **14** 1-55 (1974).

[10] A. Beauville. Journal of Diff. Geometry **18** (1983).

[11] A.N.Todorov. Moduli of Symplectic manifolds, Preprint IAS Princeton.

[12] A.N.Todorov. A description of the moduli space of Einstein metrics on a $K3$ surface. Preprint.

Currents on the torus

J. Loeffelholz
NTZ and Department of Physics
Karl Marx University
7010 Leipzig, DDR

Our goal is to study the localizability of photons. In this lecture we show the Markoff property of classical magnetostatics in R^3. We analyse in detail the case of the torus T^2 which is generic for problems showing a non-trivial topology. Then we pass to Euclidean quantum field theory.

1 Introduction

To describe the photon and its interactions with matter in the framework of constructive quantum field theory mathematicians use at least two types of continuum models. They either start directly from the Maxwell's equations [1] for the field strengthes $F_{\mu v}$, i.e.

$$F = dA \quad . \tag{1}$$

Or they work with the potentials A in some gauge [2]. For the theory of the free electromagnetic field it is well known that in the Coulomb gauge one has a Hilbert space formalism. In this case the above formula identifies the corresponding real time and Euclidean models on a rigorous level [3, 4, 5, 6, 7]. Then (1) is a unitary mapping.

But from De Rham's theory we learn that F and A induce different local structures in the mathematical model. This as we might expect must have remarkable physical consequences for problems showing non-trivial topology. We recall here the controversies concerning the existence of magnetic monopoles [8], [9], and the Aharonov-Bohm effect [10].

Finally, let us emphasize that theoreticians use also A in covariant gauges which gives a formalism with an indefinite metric [11], [12].

2 Magnetostatics

Finite energy solutions B to Maxwell's equations for classical magnetostatics [13], [14]

$$dB = 0 , \tag{2}$$
$$\delta B = J \tag{3}$$

are described by vectors in some real Hilbert space H. In this formalism a magnetic field configuration $B \in H$ is a closed square-integrable real 2-form in Euclidean space R^3. From (2) we have by the inverse Poincaré lemma $B = PM$, where

$$P = \frac{d\delta}{\Delta} \tag{4}$$

is a projection operator. Δ^{-1} stands short hand for the Green's function of the Beltrami-Laplace operator in R^3 with free boundary conditions [15], [16]. M has the physical meaning of a magnetic dipole density, and $J = \delta M$ is the 1-form of a steady electric current.

According to the localization of the sources J and M in R^3 we obtain two different local structures in H. Given an open region U, we say B belongs to the subspace $H_{(U)}$ of H if all J_i have support in the closure of that region which we denote by $\overline{U}$. And we write $B = PM \in H_U$ if

$$supp\, M_{ij} \subset \overline{U} \quad , \quad \text{for all } i, j \quad . \tag{5}$$

To construct H we could start as well from the linear space of elements M with components $M_{ij} \in C_O^\infty(R^3)$, and the inner product

$$(M, PN) \ , \tag{6}$$

which is the dipole-dipole coupling. Then H has the structure of a factor space, and the rather heuristic definition (5) means that smooth B in H_U have representants $M_{ij} \in C_O^\infty(U)$. We will denote by P_U the projection in H onto the subspace H_U. For a hypersurface S in R^3 we define H_S as the intersection of all H_U, where each U covers S. We now come to Nelson's famous Markoff property [17].

Let $B \in H_U$ and N any test source in the complement U^c of that region. We will find that all information about the magnetic field $B = PM$ with respect to U^c can be obtained by a test source on the boundary surface $S = \partial U$, i.e. $(B, N) = (B, P_S(PN))$. Hence

$$P_{U^c}B = P_S(P_{U^c}B) \text{ , for all } B \in H_U \ . \tag{7}$$

Since P appearing in the definition of the scalar product in H is a non-local projection operator the proof is tricky [18, 5, 22, 23].

The possibility of screening a magnetic field by surface currents was known already to Gauss [19] who observed that we cannot detect the true localization of the sources for the magnetic field of the earth from the outside. The standard models assume convection currents localized somewhere in the mantle of the earth [20]. However, because of the Markoff property with respect to the local structure (5) - which we are going to prove below - the origin of the field may be also dipoles frozen in the nucleus.

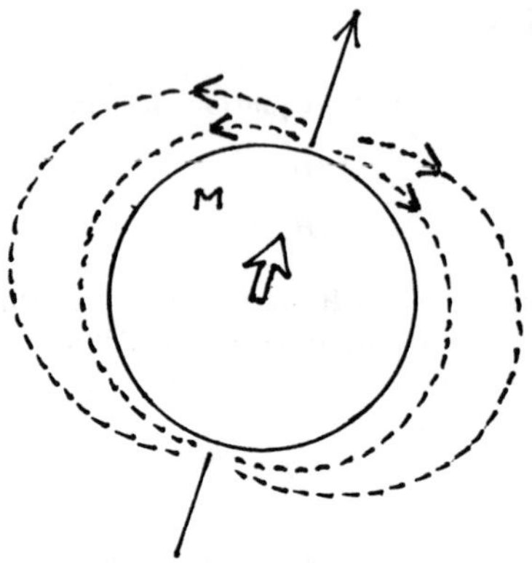

Figure 1: Compass needle N which measures $B = PM$ in U^c of R^3.

3 Topology

The Markoff property in H with respect to the local structure given by the localization of currents was shown by many authors. See e.g. [7], [24].

Let $B \in H_{(U)}$. If is simply connected like the ball of the earth - and hence U^c and S - then one can use the definition $J = \delta M$ to prove (7). Otherwise the equation $d({}^*J) = 0$ in the region U of R^3 does not imply $J = \delta M$ with $supp\, M_{ij}$ in U. Hence $B = PM$ may be not in H_U. In other words, in those cases $H_{(U)}$ is larger than H_U.

Let x_U be the characteristic function of the region U. Then $\sigma = dx_U$ has support on the boundary $S = \partial U$, and it points normal. For $J = \delta B$, $B \in H_{(S)}$, we find

$$ {}^*J = \sigma \wedge h \quad , \quad dh = 0 \quad , \tag{8} $$

where h is a 1-form defined on the two-dimensional manifold S embedded in R^3. Integrating the source equation (3), and applying Stokes' theorem we obtain for $B \in H_{(S)}$ the condition

$$ jump\,({}^*B) = h \quad \text{on } S \quad . \tag{9} $$

Let

$$ h = d\beta + \sum_{n=1}^{N} c_n h_n \tag{10} $$

be the Hodge decomposition. Then elements $h = d\beta$ give just $B = P(\beta\, {}^*\sigma)$ in H_S, and to every harmonic h_n on S there corresponds exactly one vector ω_n in the orthogonal complement $H_O = H_{(S)} \ominus H_S$.

Let U be an open m-fold connected region in R^3. Then the boundary surface $S = \partial U$ is homeomorphic to the sphere S^2 with m handles attached. The dimension of the first homology group

$$ \Pi_1(S) \tag{11} $$

is $N = 2m$.

Unfortunately, (10) does not define an othogonal decomposition in H. But in [21] we found a nice construction for all $\omega_n \in H_O$. Given U as above there are m linearly independent harmonic vector fields f_n satisfying the boundary condition

$$^*\sigma \wedge f = 0 \ . \tag{12}$$

Because of (12) which means that f_n have vanishing normal component on S they are called Neumann fields. Moreover, f_n are square-integrable in U. For those Neumann fields we define 2-forms in R^3 by

$$\omega = x_U \, ^*f \ . \tag{13}$$

From

$$d\omega = ^* \sigma \wedge f + x_U(d\, ^*f) = 0 \tag{14}$$

we conclude $\omega \in H$. And

$$^*\delta\omega = \sigma \wedge h + x_U(df) \ , \tag{15}$$

where $h = f_{/S}$ is the boundary value on $S = \partial U$, gives $\omega \in H_{(S)}$ since $df = 0$ in U. If $B = P(\beta\, ^*\sigma)$ is an arbitrary vector in H_S then (12) implies

$$(B, \omega) = \int_U \beta(\, ^*\sigma \wedge f) = 0 \ . \tag{16}$$

Hence $\omega \in H_O$. Similarly we proceed for U^c. Since elements in H are square-integrable the magnetic fields ω_n, $n = 1, 2, .., 2m$, which we obtained via (13) from the Neumann fields in U and U^c, respectively, are orthogonal to each other. Let us denote by $P_\pm$ the projection operators onto the corresponding m-dimensional subspaces $H_\pm$ in H. We have **LEMMA 1.**

$$P_+ P_- = 0. \tag{17}$$

4 The torus

For the special case $m = 1$ the hypersurface $S = \partial U$ is homeomorphic to the torus T^2 in R^3. There are two loops $l_\pm$ which one cannot contract continuously to a point on T^2. The corresponding elements $\omega_\pm \in H_O$ have the following physical meaning: The vector $\omega_+ \in H_U$ describes the magnetic field of a solenoideal current, and $\omega_- \in H_{U^c}$ is produced by the super current J_- on T^2. Of course, we have $(\omega_+ \, , \omega_-) = 0$ according to lemma 1. For a more detailed analysis let us introduce in R^3 local coordinates (r, α, φ) by

$$\begin{aligned} z &= r \sin\alpha \\ \varrho - R &= r \cos\alpha \ , \quad \text{with } r \geq 0 \text{ and } R > 1 \ . \end{aligned} \tag{18}$$

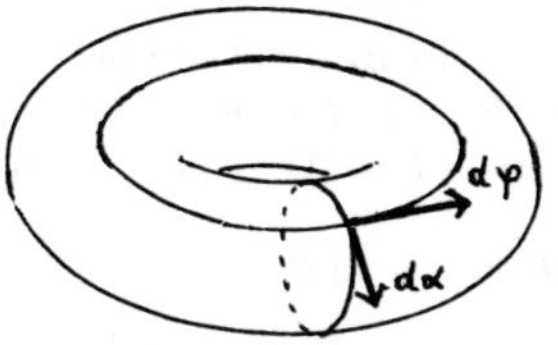

Figure 2: Local coordinates (α, φ) on T^2 in R^3, and the loops $l_\pm \in \Pi_1(T^2)$.

Above (z, ϱ, φ) are the standard cylinder coordinates. Our torus with the symmetry axis $z = 0$ is defined by the equation $r = 1$, and the restriction of the Euclidean metric to T^2 reads

$$d\alpha^2 + \varrho^2 d\varphi^2 \ , \quad \text{with } \alpha, \varphi \in S^1 \quad . \tag{19}$$

The harmonic 1-form $d\varphi$ coincides with the boundary value h_+ of the Neumann field in U. Hence ${}^*J_+ = \sigma \wedge d\varphi$. Let $\tilde{J}$ be the unit current 1-form on the z-axis. Then the associated magnetic field $\tilde{B}$ has infinite energy, i.e. $\tilde{B} \notin H$. But for the restriction to the bounded region $U = \{(r, \alpha, \varphi) : \ r < 1\}$ we get

$$\begin{aligned} P_{(U)}\tilde{B} &= x_U \, {}^*d\varphi \\ &= \omega_+ \in H_O \ . \end{aligned} \tag{20}$$

We emphasize that the boundary value $h_- \ =^* \omega_{-/T^2}$ of the Neumann field in the complement U^c is different from the harmonic 1-form $\varrho^{-1} d\alpha$.

5 Markoff property

The vectors $B = PM$ with $M_{ij} \in C_O^\infty(U)$ are dense in H_U. Hence from

$$(B, \omega_n) = \int_{U^c} M \wedge f_n = 0 \ , \quad \text{for all} \quad \omega_n \in H_- \ , \tag{21}$$

we obtain

LEMMA 2. $\quad H_{(U)} = H_U \oplus H_- \ .$

The source equation $J = \delta B$ defines a unitary mapping from H to the Hilbert space C of steady current 1-forms with the scalar product given by the Coulomb interaction energy

$$(I, \Delta^{-1} J) \ . \tag{22}$$

The inverse map to (3) is the law of Biot and Savart. Let E_U be the image of $P_{(U)}$, i.e. the projection in C onto the subspace of currents localized in $\overline{U}$. We also define $E_\pm$, etc. (22) gives the operator identity $E_U \, \delta = \delta(P_U + P_-)$, and similarly for the complement U^c. Hence with the Markoff property in C, and the above lemmas we have

$$\begin{aligned} \delta P_{U^c} P_U &= (E_{U^c} E_U - (E_+ + E_-))\delta \\ &= (E_S - E_O)\delta = \delta P_S \ . \end{aligned} \tag{23}$$

This proves (7), i.e.

LEMMA 3. The logical structure in H induced by the localization of the magnetic dipole densities satisfies the Markoff property.

Unfortunately, in a previous paper we overlooked the case of non-trivial topology. We recall that piece of our proof: Let $B = PM \in H_U$, and A an arbitrary smooth 1-form with compact support in the interior of U^c. We define $I = \delta(P_{U^c} B) = \delta N$. Then from

$$\begin{aligned} (I, A) &= (P_{U^c} B, dA) \\ &= (\delta M, A) = 0 \end{aligned} \tag{24}$$

we have $I \in C_U$, i.e. $PN \in H_{(U)}$. As remarked in section 3 without further arguments we cannot conclude that $supp\ N_{ij}$ are in $\overline{U}$.

But now, using (22) and

$$(P_{U^c}B, \omega_n) = (M, \omega_n) = 0 \quad , \text{ for all } \quad \omega \in H_- \quad , \tag{25}$$

we readily find $P_{U^c}B \in H_U$. This completes our proof given in [23].

However, we can see the Markoff property not using the currents at all. For simplicity, we again consider magnetostatics in R^3. But our method works for the model in any dimension of Euclidean space-time, and just for the standard lattice approximation. We start with

LEMMA 4. $\quad H \ominus H_U = \{\omega \in H :\ supp\,\omega_{ij} \subset U^c\}$.

Proof: Let $\omega \in H$ with $supp\,\omega_{ij} \subset U^c$. Then $(B, \omega) = 0$ for all $B = PM$ with $M_{ij} \in C_O^\infty(U)$. Since those B are dense in H_U, we conclude that $\omega \in H \ominus H_U$. Conversely, if $(M, \omega) = 0$ holds for all $M_{ij} \in C_O^\infty(U)$, then ω_{ij} should have supports in U^c. The proof is complete.

Now let $B \in H_U$. Then from

$$(P_{U^c}B, \omega) = 0 \ , \qquad \text{for all } \ \omega \ \text{ with } \ \omega_{ij} \in C_O^\infty(int\, U^c) \quad \text{satisfying } \ d\omega = 0 \tag{26}$$

we conclude $P_{U^c}B \in H_U$. Hence P_{U^c} commutes with P_U,and the proof of the Markoff property is complete.

The key was the observation of

$$H \ominus H_U \ \subset H_{U^c} \ , \tag{27}$$

which simply says that any vector in H is of the form $P(M + N)$, where M_{ij} have support in $\overline{U}$, and N_{ij} in the complement U^c. Of course, this is not an orthogonal decomposition.

A rather heuristic argument is the following: From the boundary condition

$$\sigma \wedge B = 0 \ , \quad \text{for } B \ \in H \ominus H_S \quad , \tag{28}$$

we conclude that for those vectors $(x_U + x_{U^c})B$ defines an orthogonal decomposition in H. This is non-trivial because of the condition $dB = 0$, for any $B \in H$.

According to the Hodge decomposition, for a square-integrable 2-form ω with $supp\,\omega_{ij} \subset U^c$ satisfying $d\omega = 0$ we have

$$\omega = dA + \sum_{n=1}^{m} c_n \omega_n \ , \quad \omega_n \in H_- \tag{29}$$

and $supp\, A_i \subset U^c$. We claim that those vectors span just $H \ominus H_U$. In other words, (28) unifies the two steps of our previous proof. The error of Lim in [22] was that he tried to show (28) for any 2-form η with components $\eta_{ij} \in C_O^\infty(int\ U^c)$. But

$$(B, P_{U^c}(P\eta)) = (M, P\eta) \neq 0 \ . \tag{30}$$

if P acts non-trivial.

6 Conclusions

We come back to quantum field theory. Let us recall the basic definition. The free Euclidean electromagnetic field is the Gaussian random process F_{ij} indexed by smooth 2-forms M of compact support in R^4 with mean zero and covariance

$$E\ \ F(M)F(N) = (M, PN)\ . \tag{31}$$

The above formula establishes the following correspondence to classical magnetostatics: Real one particle states of the Euclidean theory in R^3 can be identified with vectors $B \in H$.

In R^{v+1}, the magnetic field B which stands for the Maxwell tensor F_{ij} has rank two. Hence $^*B$ has the rank $v - 1$. According to (9) for a given hypersurface S, the crucial object we have to study for the Markoff property becomes the homology group

$$\Pi_{v-1}(S)\ . \tag{32}$$

The corresponding elements $B \in H_O$ are exactly the photon states which we cannot relate to an Euclidean field localized on S. See [26], [27]. They are just the ground states for the Hamiltonian $(\Delta_S + R)^{1/2}$.

For a detailed analysis of conclusion with respect to Aharonov-Bohm effect we refer to [25].

References

[1] W.Thirring, Klass.Feldth., Bd.**2**, 30, Springer (78)

[2] F.Strocchi, A.S.Wightman, J.Math.Ph.**15**, 2198 (74)

[3] P.J.M.Bongaarts, J.Math.Ph.**18**, 1510 (77)

[4] K.Osterwalder, R.Schrader, Comm.Math.Ph.**31**, 83 (73)

[5] F.Guerra, Symposia Math.Vol. **XX**, 13, Acad.Press (76)

[6] E.Seiler, Lect.Notes in Phys.**159**, Springer (82)

[7] J.Loeffelholz, Preprint KMU Leipzig, QFT **3** (74)

[8] P.A.M.Dirac, Proc.Roy.Soc. A **133**, 60 (31)

[9] M.J.Duff, J.Madore, Phys.Rev. D **18**, 2788 (78)

[10] Y.Aharonov, D.Bohm, Phys.Rev. **115**, 485 (59)

[11] K.Bleuler, Helv.Phys.Acta **23**, 567 (50)

[12] J.P.Antoine, XV.Winter School Karpacz 1978, Acta Univers.Wratisl. No **519**, 225 (79), and
J.L.Challifour, Ann.Phys. **136**, 317 (81)

[13] J.C.Maxwell, Treatise on elelctricity and magnetism, 1891, ed. Dover New York (54)

[14] J.Loeffelholz, 27.Random fields, Esztergom 1979, Coll. Math.Soc.J.Bolyai **10**, 701 (81) North Holland

[15] H.Flanders, Diff.forms and appl.. Acad.Press (63)

[16] Yu.M.Zinoviev, T.M.0.**49**, 156 (81), and **50**, 207 (82)

[17] E.Nelson, J.Func.Anal. **12**, 211 (73)

[18] T.H.Yao, J.Math.Ph. **17**, 241 (76)

[19] C.F.Gauss, Werke Bd.**V**, 1840, Gesellsch.d.Wiss.Goett.

[20] D.Rittenhouse, Inglis Rev.Mod.Phys.**53**, 481 (81)

[21] E.Martensen, Potentialtheorie, Teubner Stuttgart (68)

[22] S.C.Lim, Lett.Math.Ph. **4**, 465 (80)

[23] J.Loeffelholz, Lett.Math.Ph. **6**, 57 (82)

[24] L.Gross, Report to Cumperland Lodge Conf. (74)

[25] J.Loeffelholz, Preprint KMU Leipzig, to appear 1983

[26] A.Z.Jadczyk, B.Jancewicz Bull.Acad.Pol. **XXI**, 477 (73)

[27] A.Uhlmann, Proceed. Serpukhov Workshop (79).

Singlet Variables in Yang-Mills Theory and Matrix Models

A.A. Slavnov
Steclov Mathematical Institute,
Moscow, USSR.

In spite of great achievements in the development of gauge theories and their applications to the elementary particle physics, the problem of finding a reliable calculational scheme for large distance phenomena remains unsolved. One of the possibilities which is now under investigation is so called N^{-1} expansion. This lecture describes a new approach to this problem which was proposed in [1, 2]. The first part contains an elementary introduction to the N^{-1} expansion and its physical motivation. Then the general formalism of singlet variables for Yang-Mills theory and matrix models is described. The last sections contain applications of this formalism to the construction of N^{-1} expansion for some matrix models.

1 N^{-1} expansion. Motivation and general outline.

It is well known that the coupling constant g cannot serve as an expansion parameter for quantum chromodynamics at large distances. Renormalization group tells us that in this domain the effective coupling constant becomes large and therefore perturbation theory does not work. This fact is hardly surprising because a perturbation theory works when the spectrum of an unperturbed Hamiltonian coincides with the spectrum of the complete one. It is not the case in QCD where the unperturbed spectrum consists of the color gluons and quarks whereas the physical spectrum includes colorless mesons and baryons. The manifestation of this phenomenon is the appearence of unevitable infrared divergencies in the perturbation series over g.

The Yang-Mills Lagrangian does not contain any dimensionless parameter except for g. There is however one more parameter which enters the Lagrangian implicitly. If we shall consider the $SU(N)$ gauge group then the number N^{-1} may serve as an expansion parameter. This idea was pushed in quantum field theory by G't Hooft [3, 4] who constructed explicitly N^{-1}-expansion for two-dimensional QCD and showed that in the leading order in N^{-1} no asymptotic color states survive and the discrete series of singlet ("meson") states do exist. If these properties were present in a four dimensional case as well it would explain the phenomenon of quark confinement.

Another argument in favour of the N^{-1}-expansion is given by the phenomenology. Assuming that color is confined one can show (for details see Witten's paper [5]) that in the limit $N \to \infty$ the spectrum of singlet colorless states contains an infinite number of stable mesons, Regge phenomenology takes place, exotic states are supressed etc.

The two-dimensional models and the phenomenology both indicate that the leading order in N^{-1}-expansion enable physical properties. However the explicit construction of this expansion presents great difficulties. Let us remind the standard procedure.

The gluon field $A^j_{\mu,i}$ may be considered as an antihermitian traceless matrix $N \times N$. The propagator of this field has the following structure

$$A^i_{\mu j}(x) A^k_{\nu l}(y) = D_{\mu\nu}(x-y)\{\delta^i_l \delta^k_j - \delta^i_j \delta^k_l N^{-1}\} \quad (1)$$

The last term provides the condition $trA_\mu = 0$. and would be absent for the $U(N)$ gauge group. This term is proportional to N^{-1} and does not contribute to the leading order.

To analyze the dependence of arbitrary diagram on N it is convinient to use the following diagram technique:

The quark propagator proportional to δ^i_k is described by the single line

$$\overline{\psi^i}(x)\,\psi_k(y) = \delta^i_k S(x-y) \quad : \quad \xrightarrow{\delta^i_k} \quad . \quad (2)$$

The gluon propagator is described by the double line

$$A^i_{\mu k}(x)\,A^j_{\nu l}(y) \quad : \quad \overset{\delta^i_l}{\underset{\delta^j_k}{\overrightarrow{\overleftarrow{\quad}}}} \quad . \quad (3)$$

The quark-gluon vertex and gluon-gluon vertices look in these notation as follows

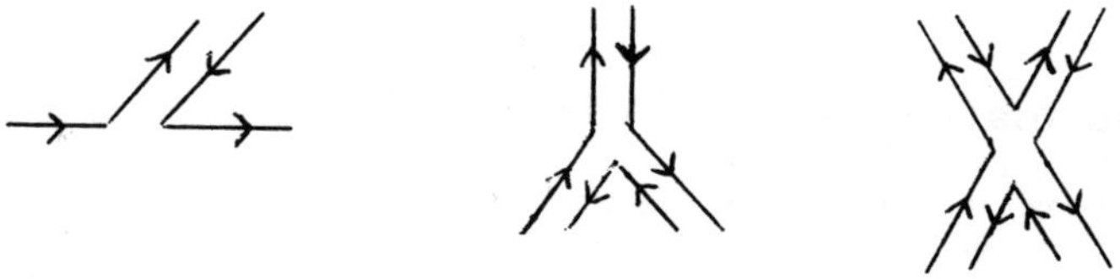

(4)

Any closed loop formed by the single lines corresponds to the trace in color indices and therefore produces the factor N.

In these notations the lowest order corrections to the gluon propagator looks as follows

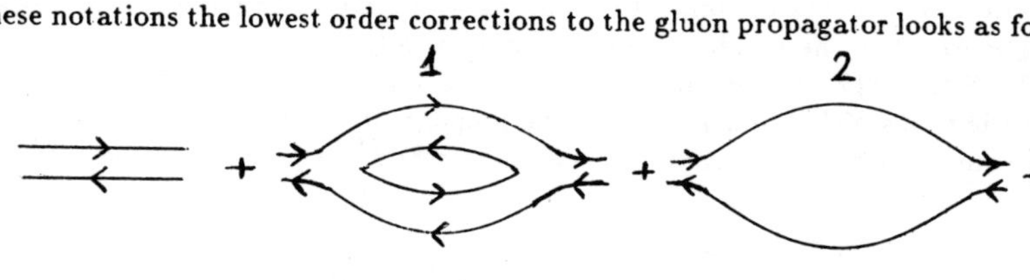

(5)

It is clear that the quark loop (2) produces the lower order term in N than the gluon loop (1). If we adopt the standard normalization $g = N^{-1/2}\tilde{g}$, where $\tilde{g}$ does not depend on N, then the diagram (1) is of zero order in N^{-1}, whereas the diagram (2) is proportional to N^{-1}.

In the same way more complicated diagrams can be analyzed, with the following result.The order in N^{-1} of arbitrary diagram is determined by the topological properties, namely the Euler characteristics of the manifold on which this diagram may be drawn without intersections. Any quark loop corresponds to the hole in the surface and decreases the order by N^{-1}. The leading order is given by the planar diagrams which may be drawn without intersections on a plane.

Therefore all the diagrams can be separated into different classes corresponding to different orders in N^{-1}, and to calculate the leading order one must sum all the planar diagrams.

The problem is indeed a hard one because the number of these diagrams is infinite and their structure in arbitrary order is quite complicated.

There are essentially two possibilities to attack the problem. One can try to derive the closed equation for the sum of planar diagrams using their topological properties. This program was realized by Yu. Makeenko and A. Migdal (see [6, 7]) who succeded to write down the closed equation for the leading order of Wilson loop

$$W(\Gamma) = \frac{1}{N} < Tr\, P \exp \int_\Gamma A_\mu(x)\, dx^\mu > \quad , \tag{6}$$

here the integral is calculated along the contour Γ. These equations however are the variational derivative equation for the contour dependent objects and at present no effective way of their solution is known.

Another possibility to avoid the difficulties of analyzing an arbitrary planar diagram is given by the reformulation of the theory in terms of singlet variables so that the number N enters the effective action only explicitly and may be used therefore as a standard perturbation expansion parameter.

2 QCD and matrix models in terms of singlet variables.

We consider the models described by the Lagrangian

$$\mathcal{L} = K_1(\varphi) + K_2(\psi) + V(\varphi) + g\, \overline{\psi}^a\, \Gamma\, \varphi_a^b\, \psi_b \quad . \tag{7}$$

Here $\varphi_a^b(x)$ are matrix field which for every x belong to the algebra $U(N)$ or $SU(N)$. They may also have Lorentz or some other indices which are not related to the "color" group $U(N)$ or $SU(N)$. These indices will be systematically supressed.

The fields $\psi^a(x)$ belong to the fundamental representation of the color group. The quadratic terms $K_1(\varphi)$ and $K_2(\psi)$ are the propagators of the fields φ and ψ respectively. We assume that these forms are not degenerate and the propagators have the following structure

$$\overbrace{\overline{\psi}^a(x)\, \psi_b(y)} = \delta_b^a\, S(x-y) \tag{8}$$

$$\overbrace{\varphi_b^a(x)\, \varphi_d^c(y)} = \delta_d^a\, \delta_b^c\, D(x-y) \quad . \tag{9}$$

The interaction $V(U)$ is a finite or infinite series of the form

$$V(\varphi) = g_3 Tr\varphi^3 + g_4 Tr\varphi^4 + \dots \, . \tag{10}$$

Evidently QCD after the fixation of the gauge falls into this calss, φ being the gluon fields and ψ being the quark fields.

If the constants g_i are normalized as follows

$$g_3 = N^{-1/2}\tilde{g}_3 \quad , \quad g_4 = N^{-1}\tilde{g}_4 \quad , \quad \dots \tag{11}$$

where $\tilde{g}_i$ does not depend on N, the leading order in N is given by the sum of the planar diagrams.

We shall show that the Green function generating functional Z, corresponding to the Lagrangian (7) may be written in the following equivalent form

$$Z = \int \exp\{ i \int [\mathcal{L}(\psi.\varphi) + s.t.]\, dx \}\, d\varphi\, d\psi = \int \exp\{ i \int [\mathcal{L}_{ef}(\zeta)\, dx + s.t.]\}\, d\zeta \quad . \tag{12}$$

Here $s.t.$ means source terms. The fields ζ are singlets with respect to the color group.

At first we consider the relatively simple case $V(\varphi) \equiv 0$. In this case the integral over φ in the left handside of the eq. (12) is Gaussian and can be calculated explicitly:

$$Z = \int \exp\{ i \int K_2(\psi)\, dx - \frac{ig^2}{2} \int (\overline{\psi}^a \psi_b)_x\, D(x-y)\, (\overline{\psi}^b \psi_a)_y\, dx\, dy + s.t. \} d\psi \quad . \tag{13}$$

(We have put for simplicity $\Gamma \equiv 1$).

The effective interaction in the formula (13) may be presented as interaction of two singlet bilocal currents via the exchange by singlet bilocal field

$$\begin{aligned} Z \;&=\; \int \exp\{ i \int K_2(\psi)\, dx + \int [\overline{\psi}^a(x)\, \psi_a(y)\, \zeta(y,x) + \\ &+\; \frac{1}{2g^2 D(x-y)}\, \zeta(x,y)\, \zeta(y,x) + \; \dots] dx\, dy \} d\overline{\varphi}\, d\psi\, d\zeta \quad . \end{aligned} \tag{14}$$

Now the integral over $\overline{\psi}$, ψ is also Gaussian and may be calculated

$$\begin{aligned} Z \;&=\; \int \exp\{ N\, Tr\, \ln[\frac{\delta^2 K_2}{\delta \overline{\psi}(x)\, \delta \psi(y)} + \zeta(y,x)] + + \\ &+\; \frac{1}{2g^2} \int \frac{1}{D(x-y)}\, \zeta(x,y)\, \zeta(y,x)\, dx\, dy + s.t. \}\, d\zeta \, . \end{aligned} \tag{15}$$

The effective action in the exponent (15) depends only on singlet bilocal fields and includes N only as a numerical factor. In the normalization $g = N^{-1/2}\tilde{g}$, N stands as a common factor at the N-independent effective action. Therefore N^{-1} expansion arises exactly in the same way as an ordinary quasiclassical expansion in $\hbar$. One should find the stationary point defined by the equation

$$\frac{i}{g^2 D(x-y)}\, \zeta(x,y) = [\frac{\delta^2 K_2}{\delta \overline{\psi}(x)\, \delta \psi(y)} + \zeta(x,y)]^{-1} = 0 \tag{16}$$

and expand the effective action near this point. The quadratic part will define the propagator of the fields ζ, and other terms will define the interaction vertices.

Any matrix model with $V(\varphi) = 0$ may be treated in this way. In particular the two-dimensional QCD, which falls into this class (in a ghost free gauge) was considered in the papers [8, 5].

Now we consider the case $V(\varphi) \neq 0$. To safe the place the explicit calculations will be done for the vertex $Tr\varphi^3$. All other vertices may be considered analogously.

The main observation is the following: the nonlinear vertices φ^n can be replaced by the vertices linear in φ with the help of new ghost fields belonging to the fundamental representation of the group under consideration. The new vertices have a structure $\overline{X}^a \varphi_a^b X_b$ completely analogous to the structure of the quark-gluan vertices we have considered above. More precisely

$$\exp\{ i \int \frac{g}{\sqrt{N}} Tr\, \varphi^3 dx \} = \int \prod_{i=1}^{3} d\overline{X}_i\, dX_i \exp\{ i \int [\frac{g^{1/3}}{N^{1/6}} [\overline{X}_1^a \varphi_a^b X_{2b} +$$
$$+ \overline{X}_2^b \varphi_b^c X_{3c} + \overline{X}_3^c \varphi_c^d X_{1d}] + i \overline{X}_1^a \theta^{-1} X_{1a} + \overline{X}_2^a X_{2a} + \overline{X}_3^a X_{3a}] dx \} . \quad (17)$$

Here θ^{-1} denotes an operator for which the inverse one posesses the properties of the θ-function at least in one argument. That means that the X_i-propagator which we shall denote symbolically as $\theta(x-y)$ one may choose for example $\prod_{\alpha=0}^{3} \theta(x_\alpha - y_\alpha)$ or $D_{ref}(x-y)$ etc. We shall assume also that some intermidiate regularization is introduced so that the φ-field propagators are not singular at $x = 0$, and also choose $\theta(0) = const \neq \infty \neq 0$. One may put $\theta(0) = 1$.

The equation (17) is checked by the explicit calculation. Integrating over X_2 and X_3 we get

$$\int d\overline{X}_1\, dX_1 \exp\{ i \int [\frac{g}{\sqrt{N}} \overline{X}_1^a \varphi_a^b \varphi_b^c \varphi_c^d X_{1d} + i \overline{X}_1^a \theta^{-1} X_1^a] dx \} =$$
$$\exp\{ Tr \ln[\delta^{ad} \theta^{-1} + \frac{ig}{\sqrt{N}} \varphi_b^a \varphi_c^b \varphi_a^c] \} = \exp\{ i \frac{g}{\sqrt{N}} \int Tr\, \varphi^3(x)\, dx \} . \quad (18)$$

The last equality follows from the well known theorem of the nonlinear analyzis. It's finite dimensional analogue is very simple: the determinant of a triangle matrix is equal to the product of the diagonal elements. From the point of view of the Feynman diagrams that means that all closed loops formed by the X_i fields are equal to zero because they contain chains of θ-function. The only nonzero term is the term proportional to $\theta(0)$. This term reproduces exactly the right hand side of the eq. (18).

Therefore we succeded to rewrite the gluon-gluon interaction in the form analogous to the quark-gluon interaction. Now we can apply to it the technique described above: integrate over the fields φ, producing the effective interaction of singlet bilocal currents, introduce singlet variables $\zeta(x,y)$ and finally integrate over the ghost fields X_i. In this way we obtain an effective action depending only on ζ and containing N only as a numerical factor. The generating functional Z may be written in the form

$$Z = \int \exp\{ N K(\zeta_i) - \frac{N^{1/3}}{g^{2/3}} C^{jk} \int \zeta_j(x,y)\, \zeta_k(y,x)\, dx\, dy + s.t. \} d\zeta_j . \quad (19)$$

Here $K(\zeta_i)$ is a functional arising after the integration over ghost fields. Explicit examples will be given below.

At first sight one may think that the integration over ζ can be performed as before using the stationary phase method and considering the second term as a small correction. It appears however that as a rule the quadratic form generated by the expansion near the stationary point is degenerate. It does not allow to use the stationary phase method in a straight forward way. Nevertheless we shall show that at least in some cases this problem can be solved and N^{-1} expansion constructed explicitly.

3 Nonlinear matrix model

As a first example we shall consider a nonlinear matrix model with the Lagrangian

$$\mathcal{L} = \frac{1}{2}\,\partial_\mu \varphi^a_b\,\partial_\mu \varphi^b_a + N^{-1}\,Tr\{\,\delta^a_b + \frac{g}{\sqrt{N}}\,\varphi^a_b\,\}^{-1} - \frac{m^2}{2}\,\varphi^a_b\,\varphi^b_a \tag{20}$$

N^{-1} expansion for this model due to the charge normalization chosen does not coincide with the topological expansion when the leading order corresponds to the sum of planar diagrams. Rather we shall sum contributions of infinite number of vertices φ^n. N^{-1} expansion for this model generates naturally some kind of superpropagator technique.

We assume that the intermediate regularization is such that the local measures, formally proportional to $\delta(0)$ are trivial (dimensional regularization for example posesses this property). This fact is not crucial but simplifies the calculations.

The Green function generating functional can be written with the help of ghost fields X in the form

$$\begin{aligned} Z &= \int \exp\{\,i\int[\,\mathcal{L}(x) + \varphi^a_b\,J^b_a\,]\,dx\,\}\,d\varphi + \\ &= \int \exp\{\,i\int[\,\frac{1}{2}\,\partial_\mu\varphi^a_b\,\partial_\mu\varphi^b_a + \varphi^a_b\,J^b_a - \frac{m^2}{2}\,\varphi^a_b\,\varphi^b_a\,]\,dx + \\ &+ i\int \overline{X}^b(x)\,[\,(\delta^a_b + \frac{g}{\sqrt{N}}\,\varphi^a_b)\,\delta(x-y) + \frac{i}{N}\,\delta^a_b\,\theta(x-y)\,]\,X_a(y)\,dx\,dy\,\}\,dX\,d\varphi \quad . \end{aligned} \tag{21}$$

Integrating over φ, introducing the singlet bilocal field $\zeta(x,y)$ and finally integrating over the ghost fields X one obtains

$$\begin{aligned} Z &= \int \exp\{\,Tr\,\ln[\,\delta^a_b\,(\delta(x-y) + \zeta(y,x) + \\ &+ \frac{i}{N}\,\theta(x-y)) - \delta(x-y)\int D(x-z)\,J^a_b(z)\,dz\,] - \\ &- \frac{i}{2}\int[\,\frac{N}{g^2}\,\frac{\zeta(x,y)\,\zeta(y,x)}{D(x-y)} + J^a_b(x)\,D(x-y)\,J^b_a(y)\,]\,dx\,dy\,\}\,d\zeta \quad . \end{aligned} \tag{22}$$

It is convenient to introduce new variables

$$\zeta(x,y) \rightarrow \zeta(x,y) - \frac{i}{N}\,\theta(y-x)\,. \tag{23}$$

Then the two-point Green function may be written as follows

$$G^{a,c}_{b,d} = \delta^a_d\,\delta^c_b\,[\,D(u-v) - \int dx\,dy\,D(y-x)\,D(v-y)\,M^{-1}(x,y)\,M^{-1}(y,x)\,\exp\{S_{ef}\}\,d\zeta \tag{24}$$

where

$$M(x,y) = \delta(x-y) + \zeta(y,x) \tag{25}$$

$$\begin{aligned} S_{ef} &= N\{\, Tr \ln[\, \delta(x-y) + \zeta(y,x)\,] - \frac{i}{2g^2} \int \frac{\zeta(x,y)\,\zeta(y,x)}{D(x-y)}\, dx\, dy\,\} + \\ &+ \frac{1}{g^2} \int \frac{\zeta(x,y)\,\theta(x-y)}{D(x-y)}\, dx\, dy \;. \end{aligned} \tag{26}$$

The last term at $N \to \infty$ is a small correction and the stationary point is defined by the equation

$$D(x-y)\,[\,\delta(x-y) + \zeta(x-y)]^{-1} - \frac{1}{g^2}\,\zeta(x-y) = 0 \tag{27}$$

or in the momentum space

$$i\,\tilde{\zeta}(p) = g^2 \int \tilde{D}(p-k)\, \frac{d^4k}{1+\zeta(k)} \quad . \tag{28}$$

This equation has an evident solution $\tilde{\zeta}(p) = const.$ For zero mass in the framework of the dimensional regularization when $D(0) = 0$, there is a unique solution $\tilde{\zeta}_{st}(p) = 0$. For nonzero mass a subtraction is needed to obtain a solution which makes sense when a regularization is removed.

Expanding the effective action near the stationary point $\tilde{\zeta}_{st} = 0$ we obtain the following quadratic form

$$\frac{N}{2}\,[\,\zeta(x,y)\,\zeta(y,x) + \frac{i}{g^2}\,\frac{\zeta(x,y)\,\zeta(y,x)}{D(x-y)}\,] \quad . \tag{29}$$

The corresponding propagator is

$$\overbrace{\zeta(x,y)\,\zeta(u,v)} = \delta(x-v)\,\delta(y-u)\,N^{-1}\,\frac{ig^2 D(x-y)}{1 - ig^2 D(x-y)} \quad . \tag{30}$$

It looks like a "superpropagator" used in the papers [9, 10, 11]. Contrary to the usual superpropagator technique the expression (31) arises necesserily in the framework of N^{-1} expansion and there exists a unique algorithm for calculation of arbitrary order in N^{-1}.

One can show that one loop correction to the Green function contains terms proportional to

$$\sim \int dx\, dy\, D(u-x)\, D(v-y)\, \frac{D^2(x-y)}{[1 - ig^2 D(x-y)]^2} \quad . \tag{31}$$

Formal expansion of this exprexssion in g^2 reproduces the ordinary perturbation theory result:

$$\sim \int dx\, dy\, D(u-x)\, D(v-y)\, D^2(x-y)\, dx\, dy \quad . \tag{32}$$

4 Bloch-Nordsiek type matrix model

As a second example we shall calculate the sum of the planar diagrams for the model described by the Lagrangian

$$\mathcal{L} = \frac{1}{2}\,\partial_\mu \varphi^a_b\, \partial_\mu \varphi^b_a + i\,\overline{\psi}^a_b\, U_\mu\, \partial_\mu \psi^b_a - m\,\overline{\psi}^a_b\, \psi^b_a + \frac{g}{\sqrt{N}}\, \overline{\psi}^a_b\, \varphi^b_c\, \psi^c_a \;. \tag{33}$$

Here the fields φ^a_b, $\overline{\psi}^a_b$, ψ^a_b for every x belong to the $U(N)$ algebra; U_μ is a constant vector, $U^2_\mu = 1$.

Such model describes heavy nonrelativistic ψ-particles interacting with the relativistic field φ. Due to the special form of the kinetic term closed loops formed from the ψ-fields are absent.

We shall calculate the ψ-field Green function

$$\begin{aligned} G'^{a,c}_{b.d} &= \frac{\delta^2 Z}{\delta J^b_a(x)\,\delta \overline{J}^d_c(y)}\Big|_{J,\overline{J}=0} \; ; \\ Z &= \int \exp\{ i \int [\mathcal{L}(x) + \overline{J}^d_c \psi^c_d - \overline{\psi}^a_b J^b_a]\, dx \}\, d\overline{\psi}\, d\psi\, d\varphi \; . \end{aligned} \tag{34}$$

Before the introduction of the ghost fields it is convenient to integrate over ψ:

$$Z = \int \exp\{ i \int [\frac{1}{2} \partial_\mu \varphi^a_b\, \partial_\mu \varphi^b_a + \overline{J}^b_a(x) [\, \delta^c_b (iU_\mu \frac{\partial}{\partial x^\mu} - m) + \frac{g}{\sqrt{N}} \varphi^c_b]^{-1} J^a_c(y)]\, dx\, dy \}\, d\varphi \; . \tag{35}$$

Having in mind that we are interested only in the terms proportional to $\overline{J}J$ we can rewrite eq. (35) with the help of ghost fields:

$$\begin{aligned} Z &= \int \exp\{ i \int [\frac{1}{2} \partial_\mu \varphi^a_b\, \partial_\mu \varphi^b_a + \overline{X}^a [\, (iU_\mu \frac{\partial}{\partial x^\mu} - m)\, \delta^b_a + \frac{g}{\sqrt{N}} \varphi^b_a]\, X_b]\, dx - \\ &- \int J^a_b(x)\, \overline{J}^c_a(y)\, \overline{X}^b(x)\, X_c(y)\, dx\, dy \}\, d\overline{X} dX\, d\varphi \; . \end{aligned} \tag{36}$$

This integral may be transformed into the integral over singlet variables in the same way as before. Finally we obtain

$$G^{a,c}_{b,d}(x,y) = i\, \delta^a_d\, \delta^c_b \int M^{-1}(x,y) \exp\{ N\, Tr\, \ln M(\zeta) - \frac{i}{2} \frac{N}{g^2} \int \frac{\zeta(x,y)\, \zeta(y,x)}{D(x-y)}\, dx\, dy \}\, d\zeta \tag{37}$$

where

$$M(\zeta) = (iU_\mu \frac{\partial}{\partial x^\mu} - m)\, \delta(x-y) + \zeta(y,x) \; . \tag{38}$$

In the leading order

$$G^{a,c}_{b,d} = \delta^a_d\, \delta^c_b\, G_o = i\, \delta^a_d\, \delta^c_b\, M^{-1}(y,x)\, |_{\zeta=\zeta_{st.}} \tag{39}$$

where $\zeta_{st.}$ is a stationary point of the effective action which is defined by the equation

$$M^{-1}(y,x) - \frac{i}{g^2} \frac{\zeta(x,y)}{D(x-y)} = 0 \; . \tag{40}$$

Therefore the Green function G_o satisfies the equation which in the momentum space looks as follows:

$$\tilde{G}^{-1}_o(p) - (up - m) - \frac{ig^2}{(2\pi)^4} \int \frac{1}{(p-k)^2} \tilde{G}(k)\, d^4k = 0 \; . \tag{41}$$

This equation may be transformed into differential one and solved exactly. G_o is determined implicitly by the integral

$$t = \int_0^{G_O^{-1}} (1 - 2\alpha \ln \frac{y}{\mu})\, dy + C \ ;$$
$$t \equiv (up) \ ; \ \alpha = \frac{g^2}{4\pi^2} \ . \qquad (42)$$

Here μ and C are some constants related to the arbitrariness in subtraction procedure.

The Green function $\tilde{G}_o(up)$ has singularities at $up = \infty$, $up = C$, and

$$\lim_{up \to C} (up - C)\, \tilde{G}_o(up) = 0 \, . \qquad (43)$$

That means the absence of a one particle pole related to the degeneracy of vacuum with respect to soft "gluons".

Analogously one can calculate the sum of so called rainbow diagrams, i.e. the diagrams without $\overline{\psi}$-loops in a model with the relativistic $\overline{\psi}$-fields.

5 Conclusions

The models considered above demonstrate that reformulation of matrix models in terms of singlet variables allows to construct effectivelly N^{-1} expansion. However at present we have no algorithm for the general case. The main problem is that in a general case the stationary phase method can not be applied straightforwardly due to the existence of zero modes. The symmetry of the effective action may help to solve this problem.

References

[1] A.A.Slavnov, Phys.Let.

[2] A.A.Slavnov, Theor.Math.Phys. **51**, 307, 1982

[3] G.'t Hooft, Nucl.Phys. **B72**, 461, 1974

[4] G.'t Hooft, Nucl.Phys. **B75**, 461, 1974

[5] E.Witten, Nucl.Phys. **B160**, 57, 1978

[6] Y.Makeenko, A.Migdal, Phys.Let. **B88**, 135, 1979

[7] Y.Makeenko, A.Migdal, Yadernaya Phys. **32**, 838, 1980

[8] D.Ebert, V.Pervuchin, Theor.Math.Phys. **36**, 313, 1978

[9] G.Efimov, JETP **48**, 598, 1965

[10] E.Fradkin, Nucl.Phys. **49**, 624, 1963

[11] H.Volkov, Ann. of Phys. **49**, 202, 1968.

III. Conformal Groups

Local Field Representations of the Conformal Group and their Physical Interpretation *

V.B.Petkova G.M.Sotkov
I.T.Todorov
Institute of Nuclear Research and Nuclear Energy,
Sofia. Bulgaria

TABLE OF CONTENTS

*Presented by I. T. Todorov.

1 Introduction: The Conformal Group of Space Time and its Universal Covering

1.1 General remarks

Elementary particle theory is becoming highly speculative.

When Einstein was searching for a unified field theory he acted as a solitary genius, isolated from the physics community. Even twenty years ago particle theorists who would venture to incorporate gravity in their consideration would be dismissed as getting old and out of pace with the current trend. Nowadays it is almost the contrary: if you do not do some "supertheory" you hardly qualify as a theorist.

The content of the present notes - the study of a class of representations of the conformal group and their field theoretic interpretation - also belongs to such a "theoretical theory": in the following pages the reader will find more mathematics than physics.

The study of conformal invariance can be regarded as part of the program of exhibiting the maximal possible symmetry of a "skeleton" 0-mass theory. (t'Hooft has argued that chiral symmetry could provide a rational for the existence of a widely different mass scales in the theory, the small masses being a trace of the broken 0-mass symmetry.)

The recognition of the conformal invariance of the vacuum Maxwell equations [B1,C3] marks probably the first application of the conformal group in physics. Our present understanding of the role of the relations $F_{\mu\nu} = \nabla_\mu A_\nu - \nabla_\nu A_\mu$ and $\nabla_\nu F^{\mu\nu} = j^\mu$ as intertwining maps between partially equivalent nondecomposable representations of the conformal group dates from recent times. It indicates the place of conformally invariant field equations in the context of the Harish Chandra theory of elementary induced representations of semi simple Lie groups [W1,D3,K2-5,R1]. It turns out that the theory of exceptional (non decomposable) representations of $G = SU(2,2)$ provides a natural framework for massless quantum electrodynamics (QED).

The present lecture reviews recent work [P6] on the structure of exceptional representations and intertwining operators of the conformal group (see also [J1],[S4],[G2],[K2-3],[R1]). We start in Sec.2 with a survey of elementary induced representations (EIRs) that contain lowest weight subrepresentations of the universal covering $\tilde{G}$ of G (see [M2],[R3],[J1]). We find it instructive to write down the x-space expression of lowest weight vectors (Sec. 2b). They exhibit the analytic properties and the asymptotic behaviour common to all representation space vectors and single out the smallest invariant subspace in the case of exeptional (reducible) elementary representations. (their knowledge also helped us to correct some inaccuracies in earlier drafts.) Lowest weight (positive energy) representations of G are precisely those which allow an analytic continuation to representations of the Euclidean conformal group Spin(5,1) (see [L1]). Thus it should not be surprising that both results and techniques developed in our earlier study of elementary representations and intertwining maps for Spin(2n+1,1) ([D2-4],[T2], see also [G1],[K1]) carry over to the present case. (This evidence is being suppressed in the otherwise interesting and instructive paper of B.Binegar et al. [B4] which also deals with conformal QED on the basis of the analysis of a class of representations of $SO(4,2)$.)

The rest of this chapter stands for a concise introduction to the conformal group and its Lie algebra. Sec 3 deals with intertwining maps and gives special attention to invariant subspace structure and partial equivalences among exceptional elementary representations that have the same values for the Casimir operators as the finite dimensional ones. (Such representations are grouped into sextets and are also called "sextet points".)

Positive energy elementary representations of the conformal group give rise to the finite component local fields used in quantum field theory (QFT) (Sec 2.). Knapp-Stein intertwining operators are interpreted as 2-points Wightman functions. Differential intertwining maps provide conformally invariant field equations. We illustrate the interrelations between the group representation and the QFT language on the simplest sextet diagram that includes the Maxwell field (and equations) in Sec. 4. (The study of another sextet diagram that involves the stress energy tensor along with the metric tensor and the Weyl curvature tensor is left as an exercise to the reader).

The Euclidean exceptional representations have proved to be crucial [D2,3] in the group theoretical derivation [M1] of the vacuum operator product expansions [B7] [F1,2]-[K6] [P7][F7]. Implications to the (Euclidean) conformal QED have been analysed in [T1,2], [K7], [P2], [F6]. The existing discussions of conformal QED (which we briefly review in Sec. 4d) lead us to believe that it might be appropriate to use a more general nondecomposable representation of the conformal group in this case. The use of such nondecomposable representations associated with the manifestly covariant formalism in 6-space has been advocated by P.Budinich and P.Furlan (see [B8 - 10] and [F8] and references therein).

1.2 The conformal group of compactified Minkowski space, its fourfold covering G = SU(2,2) and its universal covering $\tilde{G}$.

Conformal transformations are by definition angle preserving transformations.
For Minkowski space they coincide with the general (nonlinear) maps that preserve the causal orientation in the tangent space at each point. In other words, a conformal transformation of a (connected) open set U in Minkowski space M is a smooth map $x \to {}'x(x)$ of U into M satisfying

$$\eta_{\kappa\lambda}\nabla_\mu {}'x^\kappa \nabla_\nu {}'x^\lambda = \Omega^2(x)\eta_{\mu\nu} \quad (\nabla_\mu = \frac{\partial}{\partial x^\mu}) \tag{1.1}$$

where $\eta_{\mu\nu} = \mathrm{diag}(-,+,\cdots,+)$ is the metric tensor in (flat) Minkowski space and $\Omega^2 > 0$ in U. According to Liouville's theorem the conformal group of M is locally isomorphic to the (15-parameter, connected) pseudo-orthogonal group $SO_0(4,2)$ (more generally, the conformal group of 2h-dimensional Minkowski space is homomorphic for $2h > 2$ to the $(h+1)(2h+1)$- parameter group of pseudorotations $SO_0(2h,2)$). To be precise, the proper conformal group of M is the simple group

$$SO_0(4,2)/\mathbb{Z}_2 \simeq G/\mathbb{Z}_4 \simeq U(2,2)/U(1) \tag{1,2a}$$

where $U(2,2)$ is the pseudounitary group in $\mathbb{C}^4$ (i.e., in twistor space, see [T3]) and

$$G = SU(2,2) = \{g \in U(2,2); detg = 1\}. \tag{1.2b}$$

It acts as a group of fractional linear transformations of the 2×2 antihermitean matrices

$$i\underset{\sim}{x} = i\begin{pmatrix} x^0 + x^3 & x^1 - ix^2 \\ x^1 + ix^2 & x^0 - x^3 \end{pmatrix} \quad \underset{\sim}{x} = x^\mu \sigma_\mu, \sigma_0 = 1_2, \sigma_1 = \left(\begin{smallmatrix} 0 & 1 \\ 1 & 0 \end{smallmatrix}\right),\ \sigma_2 = \left(\begin{smallmatrix} 0 & -i \\ i & 0 \end{smallmatrix}\right), \sigma_3 = \left(\begin{smallmatrix} 1 & 0 \\ 0 & -1 \end{smallmatrix}\right) \tag{1.3}$$

(cf. [U1] as well as Sec. II.1a of ref [T1]).

To see that we choose a basis in $\mathbb{C}^4$ for which

$$g^*\beta q = \beta \equiv \begin{pmatrix} 0 & -1 \\ -1 & 0 \end{pmatrix} (\text{ for } g \in U(2,2)) \tag{1.4}$$

(each element of β being a multiple of the 2×2 unit matrix) and write g in a 2×2 block matrix form

$$g = \begin{pmatrix} a & b \\ c & d \end{pmatrix}. \tag{1.5}$$

Then (1.4) can be rewritten in the form

$$a^*c + c^*a = 0 = b^*d + d^*b, \; a^*d + c^*b = 1 = d^*a + b^*c, \tag{1.6}$$

under these conditions a (general) conformal transformation of x is given by

$$i\underset{\sim}{x} \xrightarrow{g} i\underset{\sim}{x'} = (ai\underset{\sim}{x} + b)(ci\underset{\sim}{x} + d)^{-1}. \tag{1.7}$$

Clearly, the centre $U(1)$ of $U(2,2)$ leaves x invariant.

It is instructive to exhibit the various subgroups of G in this realization that enter a **Bruhat decomposition** of a neighbourhood of the identity. (This decomposition will be used in Sec. 2 below to describe a physically interesting class of elementary induced representations).

Let H be the subgroup of G which leaves the point $x = 0$ invariant. H is the **parabolic subgroup** of lower triangular matrices:

$$(H \ni)h = \begin{pmatrix} \Lambda & 0 \\ i\tilde{c}\Lambda & \Lambda^{*-1} \end{pmatrix}, \tilde{c} = \tilde{c}^*, \; det\Lambda = det\Lambda^*. \tag{1.8}$$

It can be further split into the product

$$H = \Gamma_H N_4 A_1 SL(2, \mathbb{C}). \tag{1.9}$$

Here $\Gamma_H A_1 SL(2, \mathbb{C})$ corresponds to the factorization of the set of (nonsingular) 2×2 matrices Λ with real (nonzero) determinant. Identifying each factor with its image in G we have

$$SL(2, \mathbb{C}) = \{h_\Lambda = \begin{pmatrix} \Lambda & 0 \\ 0 & \Lambda^{*-1} \end{pmatrix}, det\,\lambda = 1\} \tag{1.10a}$$

(the quantum mechanical Lorentz subgroup);

$$A_1 = \{h_\alpha = \begin{pmatrix} e^{\alpha/2} & 0 \\ 0 & e^{-\alpha/2} \end{pmatrix}, \alpha \in \mathbb{R}\} \tag{1.10b}$$

(the 1-dimensional subgroup of dilations);

$$\Gamma_H = \mathbb{Z}_4/\mathbb{Z}_2 = \{\pm 1_4, \pm i 1_4\} \tag{1.10c}$$

(the factor group of the centre $\mathbb{Z}_4$ of G by the centre $\mathbb{Z}_2$ of $SL(2,\mathbb{C})$. Finally N_4 is the 4-parameter nilpotent subgroup of special conformal transformations

$$N_4 = \{h_c = \begin{pmatrix} 1 & 0 \\ i\tilde{c} & 1 \end{pmatrix}, \tilde{c} = c^\mu \tilde{\sigma}_\mu \equiv c^0 - \underline{c\sigma}, c \in M\}. \tag{1.11}$$

The Bruhat decomposition we are referring to is then written as

$$g = t_x h \qquad \text{with } t_x = \begin{pmatrix} 1 & i\underset{\sim}{x} \\ 0 & 1 \end{pmatrix}, \qquad h \in H. \tag{1.12}$$

The elements of G that do not admit such a representation belong to the 14-dimensional submanifold $wt_{x_l}H$ where x_l is a light-like vector ($x_l^2 = 0$), w is the Weyl inversion

$$w = \begin{pmatrix} 0 & \mathbf{1}_2 \\ -\mathbf{1}_2 & 0 \end{pmatrix} \tag{1.13}$$

(it commutes with $SL(2,\mathbb{C})$ and satisfies $wh_\alpha w^{-1} = h_\alpha^{-1}$).

The decomposition (1.12) allows to rederive the transformation law (1.7) if we define $({}'x =)gx$ by

$$gt_x = t_{gx}h(x,g) \quad (h(x,g) \in H)\,. \tag{1.14}$$

The special conformal transformations

$$\underset{\sim}{x} \to {}'\underset{\sim}{x} = \underset{\sim}{x}(1 - \tilde{c}\underset{\sim}{x})^{-1} = (\underset{\sim}{x} + \underset{\sim}{c}x^2)(1 + 2cx + c^2x^2)^{-1} \tag{1.15}$$

are singular on the cone $det(1 - \tilde{c}\underset{\sim}{x})(= 1 + 2cx + c^2x^2) = 0$.

Global conformal transformations are well behaved on compactified Minkowski space $\overline{M}$ which is isomorphic to the group manifold of $U(2)$(cf. [U1],[P5])

$$\overline{M} = U(2) = S^3 \times S^1/\mathbb{Z}_2 = G/H. \tag{1.16}$$

M can be embedded in $\overline{M}$ by the Cayley transform

$$i\underset{\sim}{x} \to u = (1 - i\underset{\sim}{x})(1 + i\underset{\sim}{x})^{-1} \tag{1.17a}$$

(we are using dimensionless coordinates x throughout these notes). The inverse formula

$$i\underset{\sim}{x} = (1 - u)(1 + u)^{-1} \tag{1.17b}$$

shows that there are elements of $U(2)$ that are not images of points of M; these form the 3-dimensional manifold of unitary matrices satisfying $det(1 + u) = 0$.

The manifold $\overline{M}$ is not globally causal; it contains closed time like curves of the type

$$u = e^{-i\tau}(\cos\rho - i\sin\rho\underline{n\sigma}) \;\; (\underline{n} = \sin\theta\cos\varphi, \sin\theta\sin\varphi, \cos\theta) \tag{1.18a}$$

where τ varies over a full period (for fixed ρ and $\underline{n}$ in the range $0 \le \rho \le \pi, 0 \le \theta \le \pi, 0 \le \varphi \le 2\pi, \cos\rho + \cos\tau \ge 0$). (The corresponding curve in Minkowski space

$$\underset{\sim}{x} = \frac{\sin\tau + \sin\rho\underline{n\sigma}}{\cos\tau + \cos\rho}(\tan\tau = \frac{2x^0}{1 + x^2}, \tan\rho = \frac{2\mid \underline{x} \mid}{1 - x^2}) \tag{1.18b}$$

passes through infinity for $\cos\tau + \cos\rho = 0$.

Global causality is restored on the universal covering $\tilde{M}$ of $\overline{M}$, the cylinder space

$$\widetilde{M} = \mathbb{R}^1 \times \mathbf{S}^3 = \{\tau, \cos\rho - i \sin\rho \underline{n}\underline{\sigma}; -\infty < \tau < \infty\} \tag{1.19a}$$

which has been pushed forward by Penrose and Segal (see [P5], [S2,3],[P3]). $\widetilde{M}$ is a homogeneous space of the infinite sheeted universal covering $\tilde{G}$ of G:

$$\widetilde{M} = \tilde{G}/N_4 A_1 SL(2, \mathbb{C}). \tag{1.19b}$$

The quantum mechanical conformal group should be identified with $\tilde{G}$ (since all projective representations of G are equivalent to single valued (unitary) representations of $\tilde{G}$). $\tilde{G}$ is not a matrix group. It has an infinite centre isomorphic to $\mathbb{Z} \times \mathbb{Z}_2$, where $\mathbb{Z}_2$ is the centre of $SL(2, \mathbb{C})$ and $\mathbb{Z}$ consists of all integer powers of the product ξ_1 of the Weyl inversion W with the τ -translation $\tau \to \tau - \pi$.

The space $\widetilde{M}$ can also be thought as the union

$$\widetilde{M} = \cup_{k\in\mathbb{Z}} \xi_1^k \overline{M}, \text{ where } \xi_1(\tau, \rho, \underline{n}) = (\tau - \pi, \rho - \pi, \underline{n}). \tag{1.19c}$$

We leave it as an exercise to the reader to derive

$$w t_x = t_{\underset{\sim}{x}_w} h(x, w) \tag{1.20a}$$

where

$$\underset{\sim}{x}_w = \frac{\tilde{x}}{x^2},\ h(x, w) = \begin{pmatrix} \frac{i\tilde{x}}{x^2} & 0 \\ -1 & -i\underset{\sim}{x} \end{pmatrix} = i^{\nu(x)} h_c h_\alpha h_\Lambda \tag{1.20b}$$

($\underset{\sim}{x}\tilde{x} = -x^2$) with

$$\nu(x) = -\epsilon(x^0)\theta(-x^2),\ \tilde{c}(x) = -\underset{\sim}{x},\ e^{\alpha(x)} = \frac{1}{|\, x^2 \,|},\ \Lambda(x) = \frac{i\tilde{x}}{\sqrt{|\, x^2 \,|}} i^{\nu(x)}. \tag{1.20c}$$

2 Lowest Weight Induced Representations and the Generalized Discrete Series of the Quantum Mechanical Conformal Group

2.1 Elementary induced representations of G related to local fields

We are interested in representations of $\tilde{G}$ which act in a space of (spin-tensor valued) functions on Minkowski space. We saw that it is the conformal compactification $\overline{M}$ of M (rather than M itself) that is a homogeneous space of G or $\tilde{G}$. Since, however, M is dense in $\overline{M}$ we can just speak about (smooth) functions on M obeying certain asymptotic conditions.

Thus we are lead to consider induced representations of G with inducing subgroup

$$\tilde{H} = \mathbb{Z} N_4 A_1 SL(2, \mathbb{C}) \tag{2.1}$$

satisfying

$$\overline{M} = \tilde{G}/\tilde{H} = G/H \tag{2.2}$$

(cf. (1.8),(1.16)). We consider finite dimensional irreducible representations (IRs) D_x of $\tilde{H}$. They are necessarily trivial on N_4 and are labelled by four numbers $\chi = (d; j_1, j_2; \delta)$ where δ and d give the characters of the central subgroup $\mathbb{Z}$ and the subgroup of dilations A_1, while the non-negative (half) integers j_1, j_2 refer to a finite dimensional representation of $SL(2, \mathbb{C})$ (with $2j_1$ undotted and $2j_2$ dotted indices). Each representation can be realized in the $(2j_1 + 1)(2j_2 + 1)$-dimensional space $H_{j_1 j_2}$ of homogeneous polynomials $f(\kappa, \bar{\kappa})\kappa = (\kappa_A, A = 1, 2)$ of degree of homogeneity $2j_1$ in κ and $2j_2$ in $\bar{\kappa}$ by

$$[D_\chi(\xi_1^\nu h_c h_\alpha h_\Lambda)f](\kappa, \bar{\kappa}) = \epsilon^{-\alpha d - i\pi\nu\delta} f(\kappa\Lambda, \Lambda^* \bar{\kappa}). \tag{2.3}$$

Let further C_χ be the space of infinitely smooth functions on $\tilde{G}$ with values in $\mathcal{H}_{j_1 j_2}$ which satisfy the covariance relation

$$f(gh) = D_\chi(h^{-1})f(g) \quad (g \in \tilde{G}, h \in \tilde{H}) \tag{2.4a}$$

or

$$f(g\xi_1^\nu h_c h_\alpha h_\Lambda; \kappa, \bar{\kappa}) = e^{\alpha d + i\pi\nu\delta} f(g; \kappa\Lambda^{-1}, \Lambda^{*^{-1}} \bar{\kappa}). \tag{2.4 b}$$

The elements of C_χ are in one-to-one correspondence with vector valued functions on M (satisfying certain asymptotic conditions). To see that we use the Bruhat decomposition (1.12) (with$h \in \tilde{H}$) and set

$$f(x)(\equiv f(i\underset{\sim}{x}; \kappa, \bar{\kappa})) = f(t_x). \tag{2.5}$$

The set of images f of elements of C_χ under the map (2.5) will be denoted by C_χ. The left regular representation of $\tilde{G}$ in C_χ (defined by $f(g_1) \overset{g}{\rightarrow} f(g^{-1}g_1)$) gives rise to the elementary induced representation

$$[T_\chi(g)f](x) = D_\chi(h^{-1}(x, g^{-1}))f(g^{-1}x) \tag{2.6a}$$

with $g^{-1}x$ and $h(x, g^{-1})$ determined from the Bruhat decomposition (cf. (1.14))

$$g^{-1}\begin{pmatrix} 1 & i\underset{\sim}{x} \\ 0 & 1 \end{pmatrix} = \begin{pmatrix} 1 & ig^{-1}x \\ 0 & 1 \end{pmatrix} h(x, g^{-1}) \quad (h \in \tilde{H}). \tag{2.6b}$$

We leave it to the reader as an exercise to verify that for $\delta = d + 2j_2$ (a value distinguished for positive energy representations - see eq(2.10) below) and

$$g^{-1} = \begin{pmatrix} a & b \\ c & d \end{pmatrix} \in G \quad (det(ci\underset{\sim}{x} + d) \neq 0) \tag{2.7a}$$

(2.6)can be written as

$$[T(g)f](i\underset{\sim}{x}; \kappa, \bar{\kappa}) = det(ci\underset{\sim}{x} + d)^{-d-j_1-j_2} f((ai\underset{\sim}{x} + b)(ci\underset{\sim}{x} + d)^{-1}; \kappa(d^+ - i\underset{\sim}{x}c^+), (d + ci\underset{\sim}{x})\bar{\kappa}) \tag{2.7.b}$$

while for an arbitrary $\xi_1^\nu \in \mathbb{Z}(\subset \tilde{C})$

$$T_\chi(\xi_1^\nu)f = D_\chi(\xi_1^\nu)f = e^{-i\pi\nu\delta} f. \tag{2.7c}$$

(In deriving (2.7b) we use (1.8) and note that $\Lambda^{*^{-1}} = ci\underset{\sim}{x} + d$.)

Eq (2.6) can in this case be regarded as a shorthand for (2.7).

Note on the topology of $\mathcal{C}_\chi$. Each space $\mathcal{C}_\chi$ can be equipped by a $\tilde{\kappa}$ invariant (positive) inner product

$$(f,f)_{\chi,\kappa} = \frac{8}{\pi^3}\int d^4x \mid det(1+i\underset{\sim}{x}) \mid^{2(d-2)} \mid \mathcal{D}^{(j_1 j_2)}(\Lambda(x))f(x) \mid^2 \; (\equiv \parallel f \parallel_0^2) \qquad (2.8a)$$

where $\mid f \mid^2$ is the positive SU(2)-invariant finite scalar product (e.g. for 2-component spinors $\mid f \mid^2 = \mid f_1 \mid^2 + \mid f_2 \mid^2$)

$$\Lambda(x) = (\frac{1-i\underset{\sim}{x}}{|det(1-i\underset{\sim}{x})]^{1/2}})^{-1} = \frac{e^{-i\frac{\tau}{2}} + e^{i\frac{\tau}{2}} e^{i\rho \underline{n\sigma}}}{\sqrt{2(\cos\tau + \cos\rho)}}, \qquad (2.8b)$$

$\mathcal{D}^{(j_1 j_2)}$ is the $(2j_1+1)(2j_2-1)$ dimensional irreducible representation of $SL(2,\mathbb{C})$,

$$det(1+i\underset{\sim}{x}) = 1 + x^2 + 2ix^0 = \frac{2e^{i\tau}}{\cos\tau + \cos\rho}$$
$$\mid det(1+i\underset{\sim}{x}) \mid^2 = (1+x^2)^2 + 4x^0 = \frac{4}{(\cos\tau+\cos\rho)^2}, \qquad (2.8c)$$

$$\frac{8}{\pi^3}\int d^4x \frac{1}{\mid det(1+i\underset{\sim}{x}) \mid^4} = \frac{1}{4\pi^3}\int_{-\pi}^{\pi} d\tau \int_0^{\pi} \sin^2\rho d\rho \int_0^{\pi}\int_0^{2\pi} \sin\theta d\theta d\varphi \equiv \int_{U(2)} d\mu(u) = 1 \qquad (2.8d)$$

(see (1,17-18)).

The space $\mathcal{C}_\chi$ is complete with respect to the Fréchet space topology defined by the countable set of norms

$$\parallel f \parallel_n = \parallel (1 + \sum_{a<b} J_{ab}^2)^n f \parallel_0 \quad n = 0,1,2,\cdots. \qquad (2.8e)$$

(Notice that $\sum J_{ab}^2$ is **not** the second order Casimir operator of the conformal group since all squares of generators enter with a positive sign.)

2.2 When does a positive energy subrepresentation exist?

The positive energy unitary IR's of $\tilde{G}$ have been described by Mack [M2] (see also earlier work [Y1][K6] which, essentially, also contains this clas of representations). We shall present here a slight generalization of this result that includes non-unitary lowest weight elementary representations on $\tilde{G}$ and an explicit construction of the corresponding lowest weight vectors (see also [R3]).

We start by writing down the basic infinitesimal operators of the Lie algebra of G (we adhere to the physicist convention, according to which the generators J_{ab} are hermitian for unitary representations of G). The infinitesimal operators of the elementary representations of the inducing subgroup H are

$$J_{\mu\nu} = L_{\mu\nu} + S_{\mu\nu}, L_{\mu\nu} = i(x_\nu\nabla_\mu - x_\mu\nabla_\nu), S_{\mu\nu} = \frac{1}{2}(\kappa\sigma_{\mu\nu}\frac{\partial}{\partial\kappa} - \frac{\partial}{\partial\bar{\kappa}}\sigma^*_{\mu\nu}\bar{\kappa}) \qquad (2.9a)$$

where $\sigma_{\mu\nu}$ are given by (1.37), so that $S_{12} = \frac{1}{2}(\kappa\sigma_3\frac{\partial}{\partial\kappa} - \frac{\partial}{\partial\bar{\kappa}}\sigma_3\bar{\kappa})$, $S_{03} = \frac{-i}{2}(\kappa\sigma_3\frac{\partial}{\partial\kappa} + \frac{\partial}{\partial\bar{\kappa}}\sigma_3\bar{\kappa})$

$$J_{65} = -i(d + x\nabla) = -i(d + \cos\rho\sin\tau\frac{\partial}{\partial\tau} + \sin\rho\cos\tau\frac{\partial}{\partial\rho}); \qquad (2.9b)$$

$$K_\mu = J_{\mu 6} + J_{\mu 5} = i\{2x_\mu(d + x\nabla) - x^2\nabla_\mu\} + 2x^\lambda S_{\lambda\mu}; \tag{2.9c}$$

they have to be completed by the generators of translations

$$P_\mu = J_{\mu 6} - J_{\mu 5} = -i\nabla_\mu = -i\frac{\partial}{\partial x^\mu}(\nabla_0 \equiv \frac{\partial}{\partial t} = 1 + \cos\rho\cos\tau\frac{\partial}{\partial\tau} - \sin\rho\sin\tau\frac{\partial}{\partial\rho}, etc). \tag{2.9d}$$

The starting point of ref. [M2] is the observation (made in [S2], [L1]) that energy positivity implies positivity of the compact generator (the "conformal Hamiltonian")

$$\begin{aligned} J_{60} &= \frac{1}{2}(P^0 + K^0) = i\{t(d + x\nabla) + \frac{1+x^2}{2}\nabla_0 - \frac{1}{2}(\kappa \underline{x}\underline{\sigma}\frac{\partial}{\partial\kappa} + \frac{\partial}{\partial\underline{\bar\kappa}}\underline{x\sigma\bar\kappa})\} \\ &= i\{d\frac{\sin\tau}{\cos\tau + \cos\rho} + \frac{\partial}{\partial\tau} - \frac{1}{2}\frac{\sin\rho}{\cos\tau + \cos\rho}(\kappa\underline{n\sigma}\frac{\partial}{\partial\kappa} + \frac{\partial}{\partial\underline{\bar\kappa}}\underline{n\sigma\bar\kappa})\}. \end{aligned} \tag{2.9.e}$$

(Since $P^0 \geq 0$ yields $K^0 = T_\chi(w)P^0T_\chi(w)^{-1} \geq 0$, where w is the Weyl inversion.)

Proposition: All elementary representations T of $\tilde{G}$ that are unitary when restricted to the universal cover of the maximal compact subgroup $S(U(2) \times U(2))$ of G and containing a subrepresentation for which the "conformal Hamiltonian" J_{60} is bounded below, are of the type

$$\chi = (d; j_1; j_2; \delta)\text{with } d = \bar{d}(\text{real }) \text{ and} \delta = d + 2j_2(\text{mod}2) \tag{2.10}$$

and min $J_{60} = d$.

sketch of the proof. Since $e^-i2\pi J_{60}(= T_\chi(\xi_1^2))$ belongs to $\mathbb{Z}$ (hence to the centre of the elementary representation T_χ) eq. (2.3) implies that the spectrum of J_{60} is discrete: it belongs to the set $\{\delta + n\}$ where n is an integer. A ssuming that it is bounded below, let ω_0 be its minimal eigenvalue. If

$$J_{60}f = \omega_0 f \tag{2.11}$$

then

$$(J_{a0} + iJ_{6a})f = 0 \text{ for } a = 1,2,3,4,5, \tag{2.12}$$

since the commutation relation $[J_{60}, J_{a0} + iJ_{6a}] = -J_{a0} - iJ_{6a}$ would otherwise imply that $J_{60}(J_{a0} + iJ_{6a})f = (\omega_0 - 1)(J_{a0} + iJ_{6a})f$ contrary to our assumption.

A lowest weight state $f_{dj_1j_2}$ satisfying (2.11) (2.12) and

$$(J_{12}^2 + J_{13}^2 + J_{23}^2)f_{dj_1j_2} = s_0(s_0+1)f_{dj_1j_2}, (J_{12} + s_0)f_{dj_1j_2} = 0 \tag{2.13}$$

where $s_0 =| j_1 - j_2 |$, only exists for $\omega_0 = d$ and then it is unique. In order to find it, it is useful to display the lowering operators in the form

$$\begin{aligned} J_{50} + iJ_{65} &= d(1 + it) + e^{i\tau}(\sin\rho\frac{\partial}{\partial\rho} - i\cos\rho\frac{\partial}{\partial\tau}) + \\ &+ \frac{1}{2i}(\kappa\underline{\sigma x}\frac{\partial}{\partial\kappa} + \frac{\partial}{\partial\bar\kappa}\underline{\sigma x}\bar\kappa) \end{aligned} \tag{2.14a}$$

$$\begin{aligned} J_{30} + iJ_{63} &= n_3[dr - e^{i\tau}(\cos\rho\frac{\partial}{\partial\rho} + i\sin\rho\frac{\partial}{\partial\tau})] + e^{i\tau}\frac{\sin\theta}{\sin\rho}\frac{\partial}{\partial\theta} + \frac{i-t}{2}(\kappa\sigma_3\frac{\partial}{\partial\kappa} + \\ &+ \frac{\partial}{\partial\bar\kappa}\sigma_3\bar\kappa) + \frac{i}{2}[\kappa(\underline{x}\wedge\underline{\sigma})_3\frac{\partial}{\partial\kappa} - \frac{\partial}{\partial\bar\kappa}(\underline{x}\wedge\underline{\sigma})_3\bar\kappa] \end{aligned} \tag{2.14b}$$

and similar expressions for $a = 1, 2$; here $(\underline{x} \wedge \underline{\sigma})_3 = x_1\sigma_2 - x_2\sigma_1$, and notations are as in (1.18).

A straightforward calculation gives

$$f_{dj_1j_2}(i\underline{x};\kappa,\overline{\kappa}) = \frac{N[2(i-t)\kappa\underline{x}\overline{\kappa} + (1-x^2)\kappa\overline{\kappa}]^{2j_{min}}}{(1+x^2+2it)^{d+j_1+j_2}} \begin{cases} (\kappa(1+i\underline{x})\,|_2)^{2j_1-2j_2} & j_1 \geq j_2 \\ ((1+i\underline{x})\overline{\kappa}\,|_2)^{2j_2-2j_1} & j_2 > j_1 \end{cases} \tag{2.15}$$

where N is a normalization constant, and $j_{\min} = \min(j_1, j_2)$,

$\kappa(1+i\underline{x})\,|_2 = \kappa_2[1+i(x^0-x^3)] + \kappa_1(ix^1+x^2), (1+i\underline{x})\overline{\kappa}\,|_2 = [1+i(x^0-x^3)]\overline{\kappa}_2 + (ix_1-x_2)\overline{\kappa}_1$.

In deriving (2.15) we have used the relations

$$1 + x^2 + 2it = \frac{2e^{i\tau}}{\cos\tau + \cos\rho}, (1+i\underline{x})^2 = 2(i-t)\underline{x} + 1 - x^2. \tag{2.16}$$

Combining the expression (1.20) for the action of the Weyl inversion with the relation

$$T_\chi(\xi_1) = e^{-i\pi J_{60}} T_\chi(w) \tag{2.17a}$$

and with the definition (2.3) of D_χ we find

$$T_\chi(\xi_1) f_{dj_1j_2}(x) \equiv e^{-i\pi\delta} f_{dj_1j_2}(x) = e^{-i\pi(d+2j_2)} f_{dj_1j_2}(x). \tag{2.17b}$$

(In deriving the right hand side of the last equation we used the explicit form (2.15) of $f_{dj_1j_2}$ and the identity $i\underline{x}(1+i\underline{x}_w) = 1 + i\underline{x}$.) Thus $e^{i\pi(d+2j_2-\delta)} = 1$, which implies (2.10).

Remark Similarly negative energy (highest weight) elementary representations can be treated; for them

$$\delta = -d + 2_{j_1}(\text{mod}2). \tag{2.18}$$

3 Partial equivalences.

3.1 Introduction and review

The conformal Casimir operators are expressed as follows in terms of the labels j_1, j_2 and $c = d - 2$ of the elementary induced representations [Y1]

$$C_2 \equiv \frac{1}{2} J_{ab} J^{ab} = 2 j_1 (j_1 + 1) + 2 j_2 (j_2 + 1) + c^2 - 4 \tag{3.1a}$$

$$C_3 \equiv \frac{1}{2^3 3!} \varepsilon_{abcdef} J^{ab} J^{cd} J^{ef} = (j_1 - j_2) (j_1 + j_2 + 1) c \tag{3.1b}$$

$$C_4 \equiv J_{ab} J^{bc} J_{cd} J^{da} - C_2^2 - 8 C_2 = (2 j_1 + 1)^2 (2 j_2 + 1)^2 - \tag{3.1c}$$
$$- 2[(j_1 + j_2 + 1)^2 + (j_1 - j_2)^2] + 1 - \{2[(j_1 + j_2 + 1)^2 + (j_1 - j_2)^2 + 1] - c^2\}c^2 .$$

Considering general elementary representations of G we have to add to this list the central element

$$T_\chi (\zeta_1) = e^{-i\pi\delta} \tag{3.2}$$

(from which δ is determined modulo 2). We observe that all these invariant operators have the same values for dual representations

$$\chi = (2 + c (= d) ; j_1 , j_2 ; \delta) , \quad \tilde{\chi} = (2 - c ; j_2 , j_1 ; \delta (mod 2)) . \tag{3.3}$$

In the generic case such dual representations are equivalent and (topologically) irreducible (i.e. they do not have nontrivial closed invariant subspaces). For

$$\delta = d + 2 j_2 (mod 2) \quad (or \quad \delta = -d + 2 j_1 (mod 2)) \tag{3.4}$$

T_χ admits as we saw a lowest (or highest) weight subrepresentation, and, hence, it is reducible. The Knapp-Stein intertwining maps (to be constructed below) only establish partial equivalence [1] between T_χ and $T_{\tilde{\chi}}$ (it could not be otherwise, because, if χ admits a lowest weight subrepresentation, then, for non-integer $d + j_1 + j_2$, $\tilde{\chi}$ only admits a highest weight subrepresentation and vice versa).

For integer $d + j_1 + j_2$ (we shall call such χ's integer points for short) the structure of invariant subspaces and intertwining maps is richer (and more complex). We shall be mainly interested in a class of exceptional points for which (3.4) is also satisfied. We first note that

$$c + 2 j_2 = - c + 2 j_1 (mod 2) \quad \text{for integer } c + j_1 + j_2 \tag{3.5}$$

so that such representations involve both positive and negative energy invariant subspaces. Exceptional representations for which

$$max (c + 2 j_2 , 2 j_1 - c) \geq 2 + min (| j_1 - j_2 | , || j_1 - j_2 | -1 |) \tag{3.6}$$

can be grouped in sextets of partially equivalent representations with the same values of the Casimir operators. Each sextet can be labelled by a positive integer ν, by a non-negative l for which $l + j_1 + j_2$ is an integer , and, by a (natural) number n that can take the values $1, ..., 2l + 1$, in the way shown on Fig.1 (where the short hand notation C_d instead of $C_{(d; j_1, j_2; d+2j_2)}$ is being used for the representation spaces within a sextet).

[1] A map $W : C_{\tilde{\chi}} \to C_\chi$ is called intertwining if $T_\chi(g) W = W T_{\tilde{\chi}}(g)$. An intertwining map establishes (complete) equivalence between T_χ and $T_{\tilde{\chi}}$ iff it is invertible (so that $T_\chi = W T_\chi W^{-1}$) otherwise it only gives rise to a partial equivalence.

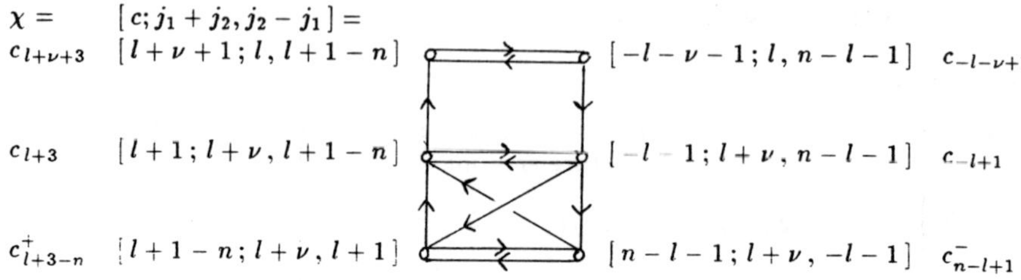

Figure 1: Sextet of exceptional points in the space of H-induced elementary representations. Arrows indicate intertwining maps which will be displayed in the following subsections.

The range of the parameters l , ν , n on Fig.1 is

$$l = 0, \frac{1}{2}, 1, ... ; \quad \nu = 1, 2, ... , \quad n = 1, 2, ..., 2l + 1 \ . \tag{3.7}$$

(Notice that the representation space $C_{-l-\nu+1}$ contains for all l and ν a finite dimentional invariant subspace.) Observe further that all representations of the type pictured on Fig.1 are single valued representations of $G = SU(2,2)$.

Although the triplets (c, j_1, j_2) are different in each of the three points of a given column the combinations

left column $\quad c + 2j_2 = 3l + \nu + 2 - n$
right column $\quad -c + 2j_1 = 3l + \nu + 2 - n$

are always the same. We leave it as an exercise to the reader to verify that the Casimir invariants for all six representations of Fig.1 are the same and hence coincide with their values for the finite dimensional representations.

3.2 Knapp-Stein intertwining operators and invariant forms

According to the general algorithm [K4] (see also [D3] where the Knapp-Stein construction has been sysytematically applied to elementary induced representations of the Euclidean conformal group) the intertwining map $W_\chi : C_{\tilde{\chi}} \to C_\chi$ is expressed in terms of the Weyl inversion:

$$W_\chi f(g; \kappa; \bar{\kappa}) = \gamma(\chi) \int_{\mathbb{R}^4} d^4x\, f(g w t_x; \bar{\kappa}\varepsilon, {}^t\varepsilon\kappa,) \tag{3.8a}$$

where

$$\varepsilon = i\sigma_2 (= -{}^t\varepsilon = -\varepsilon^{-1}) \text{ satisfies } \varepsilon \overset{t-1}{\Lambda} \varepsilon^{-1} = \Lambda \ , \tag{3.8b}$$

t_x (1.12) runs over the subgroup of translations and $\gamma(\chi)$ is a normalization factor. As a consequence the operator W_χ defines an invariant sesquilinear form on the space $C_{\tilde{\chi}}$:

$$(f_1, f_2) = \frac{1}{(2j_1)!(2j_2)!} \int d^4x\, \bar{f}_1(x; \frac{\partial}{\partial\kappa}{}^t\varepsilon, \varepsilon\frac{\partial}{\partial\bar{\kappa}})(W_\chi f_2)(x; \kappa, \bar{\kappa}) \tag{3.9}$$

for $f_1, f_2 \in C_{\tilde{\chi}}$.
If T_χ contains a lowest weihgt subrepresentation, then (3.8) can be rewritten in the x-space

picture as

$$[W_\chi f](i\underline{x}_1;\kappa^1,\overline{\kappa}^1) = \tag{3.10a}$$
$$= \frac{1}{(2j_1)!\,(2j_2)!}\int d^4x_2\, W_\chi(x_{12};\kappa^1,\overline{\kappa}^1;\frac{\partial}{\partial\kappa^2}{}^t\varepsilon,\varepsilon\frac{\partial}{\partial\overline{\kappa}^2})\,f(i\,x_2,\kappa^2,\overline{\kappa}^2)$$

where $x_{12} = x_1 - x_2$,

$$W_\chi(x;\kappa^1,\overline{\kappa}^1:\kappa^2,\overline{\kappa}^2) = \tag{3.10b}$$
$$= \frac{n(\chi)}{(2\pi)^2}\left(\frac{\kappa^1\, i\,\underline{x}\,\overline{\kappa}^2}{\sqrt{2}}\right)^{2j_1}\left(\frac{\kappa^2\, i\,\underline{x}\,\overline{\kappa}^1}{\sqrt{2}}\right)^{2j_2}\left(\frac{2}{x^2+i\,o\,x^o}\right)^{d+j_1+j_2}.$$

The normalization constant $n(\chi)$ can be choosen in such a way that $W_\chi \circ W_{\tilde{\chi}} = I_\chi$ (the identity operator on $\mathcal{A}_\chi$)

$$\int d^4y\, W_\chi(x_1-y;\kappa^1,\overline{\kappa}^1;\frac{\partial}{\partial\kappa}{}^t\varepsilon,\varepsilon\frac{\partial}{\partial\overline{\kappa}})\,W_{\tilde{\chi}}(y-x_2;\kappa,\overline{\kappa};\kappa^2,\overline{\kappa}^2) =$$
$$= 2\pi\int d_4p\,\theta_+(p)\,e^{ipx_{12}}(\kappa^2\varepsilon\kappa^1)^{2j_1}(\overline{\kappa}^2\varepsilon\overline{\kappa}^1)^{2j_2}\quad (d_4p = \frac{d^4p}{(2\pi)^4}) \tag{3.11}$$

where $\theta_+(p) = \theta(p^o)\,\theta(-p^2)$. Here $W_{\tilde{\chi}}$ is given by (3.10b) with (d, j_1, j_2) replaced by $(4-d, j_2, j_1)$.
The right hand side of (3.11) is just the kernel of the projection operator on the positive energy symmetric tensors of $2j_1$ indices.

For non integer $d + j_1 + j_2$, if χ contains a lowest weight subrepresentation (acting in$\mathcal{A}_\chi \subset C_\chi$), then $\tilde{\chi}$ contains a highest weight subrepresentation (acting in $\overline{\mathcal{A}}_{\tilde{\chi}} \subset C_{\tilde{\chi}}$). The intertwining operator W_χ (3.10) establishes equivalence between the factor representation of $T_{\tilde{\chi}}$ in $C_{\tilde{\chi}}/KerW_\chi$ ($KerW_\chi \supset \overline{\mathcal{A}}_{\tilde{\chi}}$) and the subrepresentation of T_χ in $\mathcal{A}_\chi$. In other words W_χ maps $C_{\tilde{\chi}}$ into $\mathcal{A}_\chi$ and annihilates $\overline{\mathcal{A}}_{\tilde{\chi}}$:

$$W_\chi : C_{\tilde{\chi}} \to C_\chi,\ Im\,W_\chi \subset \mathcal{A}_\chi,\ Ker\,W_\chi \supset \overline{\mathcal{A}}_{\tilde{\chi}},\ W_\chi T_{\tilde{\chi}} = T_\chi W_\chi. \tag{3.12}$$

Note that the normalization condition (3.11) cannot be interpreted as a superposition of conformal maps. Indeed, the image of W_χ as defined by (3.10) is always a positive energy ($\mathcal{A}_\chi$ -type) space. But if $C_\chi \supset \mathcal{A}_\chi$ (that is, if $\delta = c + 2j_2\,(mod\,2)$) then $W_{\tilde{\chi}}$ is not a conformal map of C_χ into $C_{\tilde{\chi}}$ (since $C_{\tilde{\chi}}$ contains no conformal invariant positive energy subspace for noninteger $c + j_1 + j_2$). Nevertheless, $W_{\tilde{\chi}}$ defines a conformally invariant hermitian form on $\mathcal{A}_\chi$ given by

$$\int dx_1\int dx_2\,\overline{f}_1(x_1;\frac{\partial}{\partial\overline{\kappa}^1}{}^t\varepsilon,\varepsilon\frac{\partial}{\partial\kappa^1})\,W_{\tilde{\chi}}(x_{12};\kappa^1,\overline{\kappa}^1;\frac{\partial}{\partial\kappa^2}{}^t\varepsilon,\varepsilon\frac{\partial}{\partial\overline{\kappa}^2})\,f_2(x_2;\kappa^2,\overline{\kappa}^2) \tag{3.13}$$

for $f_1, f_2 \in \mathcal{A}_\chi$. It is connected to the form (3.9) on $C_{\tilde{\chi}}/Ker\,W_\chi$ (with W_χ given by (3.12)) as a result of the normalization condition (3.11). The form (3.9) can be also rewritten as

$$(f_1, f_2) = \frac{1}{(2j_1)!\,(2j_2)}\int dx\,\widehat{f}_1(x;\frac{\partial}{\partial\kappa}{}^t\varepsilon,\varepsilon\frac{\partial}{\partial\overline{\kappa}})\,\varphi_2(x;\kappa,\overline{\kappa}) \tag{3.9'}$$

where $\varphi_2 = W_\chi f_2$ belongs to $\mathcal{A}_\chi$ if T_χ contains a lowest weight subrepresentation and $\hat{f}_1 = \overline{f}_1$ transforms according to the contragradient representation $\widehat{\chi} = (2-c\,;\, j_1, j_2\,;\, -\delta)$. Eq. (3.9') can be used to define an extension of the space $\mathcal{A}_\chi$ (and of the corresponding representation $T_\chi|_{\mathcal{A}_\chi}$) to the space of covariant linear functionals on $C_{\widetilde{\chi}}$.

We shall only illustrate the peculiarities of the Knapp-Stein operators at integer points by considering two simple examples: one in which the inequality (3.6) is violated (and we again have a pair of partially equivalent elementary representations) and another belonging to a sextet of the type displayed on Fig.1.

As an example of a doublet-type integer point we take the scalar field representation χ_s with $d = 1\,(=\delta)$, $j_1 = j_2 = 0$. The kernel (3.10b) coincides in this case with the 2-point Wightman function of a free 0-mass field. Taking the normalization constant $n(\chi_s) = \frac{1}{2}$ and using the notation $W_{\chi_s} = W_S$ (for $\chi_s = (1,0,0)$), we find

$$W_S(x) = \frac{1}{(2\pi)^2}\,\frac{1}{x^2 + i\,o\,x^o} = 2\pi \int \delta_o^+(p)\, e^{ipx} d_4p = \int_{C_{3,1}^+} e^{ipx} (dp)_o \tag{3.14a}$$

where

$$\delta_m^+(p) = \theta(p^o)\,\delta\,(m^2 + p^2)\,,\; C_{3,1}^+ = \{p \in M\,;\; p^o = |\,\underline{p}\,|\}\,,\; (\,dp\,)_o = \frac{d^3p}{(2\pi)^3\, 2\,|\,\underline{p}\,|}\;. \tag{3.14b}$$

We see that the image of $C_{\tilde{\chi}_S}$, $\tilde{\chi}_S = (3;0,0)$ under the map (3.10)

$$\varphi(x_1) = \int W_S(x_{12})\, f(x_2)\, d^4x_2 \qquad (\,f \in C_{\tilde{\chi}_s}\,,\, \varphi \in \mathcal{A}_{\chi_s} \equiv \mathcal{A}_S\,) \tag{3.15}$$

is only a subspace of $\mathcal{A}_S$: the space $\mathcal{A}$ of (positive energy) solutions of the d'Alembert equation

$$\Box\, \varphi(x) = 0\,,\; \Box = \nabla_\mu \nabla^\mu = \Delta - \frac{\partial^2}{\partial t^2}\;. \tag{3.16}$$

The subspace $\mathcal{A}$ of $\mathcal{A}_S$ is conformally invariant so that the representation χ_s is (further) reducible. We stress that the wave equation (3.16) (which is always Poincare invariant) is only conformally invariant for scale dimension $d = 1$.

On the other hand, the map $W_{\tilde{S}} (\equiv W_{\tilde{\chi}_s}) \,:\, C_{\chi_s} \to C_{\tilde{\chi}_s}$ with kernel

$$W_{\tilde{S}}(x) = \frac{1}{\pi^2}\,(\frac{2}{x^2 + i\,o\,x^o})^3 = 2\pi \int \theta_+(p)\,(-p^2)\, e^{ipx} d_4\,p \tag{3.17}$$

reduces to the d'Alembert operator when restricted to $\mathcal{A}_S$ and hence annihilates $\mathcal{A}$. Thus

$$W_{\tilde{S}} \,*\, W_S \equiv \int W_{\tilde{S}}(x_1 - y)\, W_S\,(y - x_2)\, d^4y = 0 \;\; (= W_S \,*\, W_{\tilde{S}}) \tag{3.18}$$

so that the normalization condition (3.11) cannot be realized. Then we can still define a conformal hermitian form on the invariant subspace $\mathcal{A}_{\tilde{S}}$ of $C_{\tilde{\chi}_s}$ obtained as the image of $W_{\tilde{S}}$:

$$\mathcal{A}_{\tilde{S}} = \{\, f = \Box\, \varphi\,,\, \varphi \in \mathcal{A}_S \,\}\;. \tag{3.19}$$

Such a role is played by the map B_S with kernel

$$B_S(x) = \frac{-ln\,[\,(x^2 + i\,o\,x^o)\,\mu^2\,]}{4\pi^2\,(x^2 + i\,o\,x^o)} = \frac{d}{d\varepsilon}\,\frac{[\mu^2\,(x^2 + i\,o\,x^o)\,]^{-\varepsilon}}{4\pi^2\,(x^2 + i\,o\,x^o)}\,\Big|_{\varepsilon\,=\,0} \tag{3.20}$$

where μ^2 is an arbitrary (positive) mass parameter.

The form

$$\int dx_1 \int dx_2\, \overline{f}(x_1)\, B_S(x_{12})\, f(x_2) = (f\,,\, B_S\, f) \tag{3.21}$$

is invariant and μ independent for $f \in \mathcal{A}_{\tilde{S}}$, although it is neither on the entire space $C_{\tilde{\chi}_s}$. (Moreover, unlike $(f, W_S f)$ it is nontrivial on $\mathcal{A}_{\tilde{S}}$.)

As a sextet-type example we consider the vector field representation χ_v with $d = 1$, $j_1 = j_2 = 1/2\,(\delta = 0)$; it belongs to the sextet with $l = 0$, $\nu = n = 1$. The W-kernel is longitudinal in this case:

$$\begin{aligned} W_V(x_{12}; k_1, k_2) &= \frac{k_1\, r(x_{12})\, k_2}{8\pi^2\,(x_{12}^2 + i\,o\,x_{12}^o)} = \frac{k_1\, \nabla_{12}\, k_2\, \nabla_{12}}{(4\pi)^2}\, ln\, \mu^2\,(x_{12}^2 + i\,o\,x_{12}^o) \\ &= 2\pi \int k_1 p\, k_2 p\, \delta'(-p^2)\, \theta\,(p^o)\, e^{ipx_{12}}\, d_4\, p \end{aligned} \tag{3.22}$$

where

$$\begin{aligned} k_i p &= \frac{1}{\sqrt{2}}\, \kappa_i\, \underset{\sim}{p}\, \overline{\kappa}_i \quad (i = 1, 2) \;\; \nabla_{12} = \frac{\partial}{\partial x_{12}} \\ k_1\, r\,(x_{12})\, k_2 &= k_1\, k_2 - 2\, \frac{k_1\, x_{12}\;\, k_2\, x_{12}}{x_{12}^2 + i\,o\,x_{12}^o}\;\; . \end{aligned} \tag{3.23}$$

(As clear from the first equation (3.22) W_V is actually independent of the mass parameter μ.) Therefore, W_V maps $C_{\tilde{\chi}_v}$ onto an invariant subspace $\mathcal{A}_V^{o\, long}$ of longitudinal (positive energy) vector fields l_μ satisfying

$$\mathcal{A}_V^{o\; long} = \{\, l_\mu^{(o)}\,(x) \in \mathcal{A}_V\;;\; l_\mu^{(o)}(x) = \nabla_\mu\, s(x)\;,\; \Box^2\, s(x) = 0\,\}\;\; . \tag{3.24}$$

Again in this case $W_{\tilde{V}}$ is equivalent to a differential operator on $\mathcal{A}_V$:

$$W_{\tilde{V}}\,(x\,;\,k_1, k_2) = \frac{3}{2\pi^2}\,(\frac{2}{x^2 + i\,o\,x^o})^3\, k_1\, r(x)\, k_2 = 2\pi \int \theta_+(p)\,(k_1\, k_2\,(-p^2) + p\, k_1\; p\, k_2)\, e^{ipx}\, d_4 p\;\; . \tag{3.25}$$

Indeed W_V maps $\mathcal{A}_V$ into the subspace C_J^{tr} of traverse currents

$$\mathcal{A}_J^{Mxw} = W_{\tilde{V}}\, \mathcal{A}_V = \{\, j_\mu = -\Box\, \alpha_\mu + \nabla_\mu\, \nabla^\alpha\;,\; \alpha \in \mathcal{A}_V\,\} \subset C_J^{tr} = \{\, j_\mu \in \mathcal{A}_{\tilde{V}}\;;\; \nabla\, j = 0\,\}\;\; . \tag{3.26}$$

The set $\mathcal{A}_V \cap Ker\, W_{\tilde{V}}$ of all elements of $\mathcal{A}_V$ which are annihilated by $W_{\tilde{V}}$ form a bigger invariant subspace of $\mathcal{A}_V$ that includes all longitudinal vector fields

$$\mathcal{A}_V^{long} = \{\, l_\mu(x) = \nabla_\mu\, s(x)\;,\; s(x) \in \mathcal{A}_{\chi_o}\;,\; \chi_o = (0; 0, 0)\,\} \supset \mathcal{A}_V^{o\; long} \tag{3.27}$$

as well as all solutions of the free Maxwell equation with a conformal gauge fixing (CGF) (studied in [M5] [B2] [S6, 7]):

$$\mathcal{A}_V^{CGF} = \{\alpha \in \mathcal{A}_V\;;\; \Box\, \alpha_\mu - \nabla_\mu\, \nabla\alpha = 0 = \Box\, \nabla\, \alpha\}\;\; . \tag{3.28}$$

It turns out that the set of all (positive energy) solutions of the free Maxwell equations is the union of (3.27) and (3.28) :

$$\mathcal{A}_V^{free} \equiv \{\, \alpha \in \mathcal{A}_V\;;\; \Box\, \alpha_\mu - \nabla_\mu\, \nabla\, \alpha = 0\,\} = \mathcal{A}_V^{long} \cup \mathcal{A}_V^{CGF} \tag{3.29}$$

while

$$\mathcal{A}_V^{long} \cup \mathcal{A}_V^{CGF} = \mathcal{A}_V^{o\,long} \quad . \tag{3.30}$$

To sumarize, the picture of invariant subspaces in $\mathcal{A}_V$ ($\subset C_{\chi_V}$) is given by

$$\mathcal{A}_V^{o\,long} \begin{array}{c} \subset \mathcal{A}_V^{CGF} \subset \\ \subset \mathcal{A}_V^{long} \subset \end{array} \mathcal{A}_V^{free} \subset \mathcal{A}_V \subset C_{(1;\frac{1}{2},\frac{1}{2};0)} \quad . \tag{3.31}$$

To the more complex structure of subrepresentations corresponds a richer family of intertwining maps (compared with doublet integer points like χ_s). For example, the Wightman 2-points function for a free electromagnetic potential

$$W_{\mu\nu}(x;\zeta) = 2\pi \int \theta(p^o)\,(p^2\,\eta_{\mu\nu} - (1-\zeta)\,p_\mu p_\nu\,)\,\delta'(-p^2)\,e^{ipx} d_4 p \tag{3.32a}$$

$$= \frac{1}{(4\pi)^2}\,(\,\Box\,\eta_{\mu\nu} - (1-\zeta)\,\nabla_\mu \nabla_\nu\,)\,ln\mu^2\,(x^2 + i\,o\,x^o) \tag{3.32b}$$

(where ζ is an arbitrary gauge parameter) defines a conformal intertwining map form the space C_J^{tr} ($\subset C_{\widetilde{V}}$ of conserved currents into $\mathcal{A}_V^{CGF}$ that is expressed in terms of the scalar function (3.14)

$$W_{\mu\nu} : (\,\mathcal{A}^{tr} \ni)\,j_\mu(x_1) \to \int W_S(x_{12})\,j_\mu(x_2)\,d^4x_2 = 2\pi \int \delta_o^+(p)\,\widetilde{j}_\mu(p)\,e^{ipx_1}\,d_4 p \in \mathcal{A}_V^{CGF} \quad , \tag{3.33}$$

and does not dependent on the gauge parameter ζ. (It plays the role of a B-kernel in the terminology used in the description of the (non-exceptional) scalar case.)

Another less familiar example is given by the invariant hermitian form in the space of Maxwell currents (see (3.26))

$$(\,j^\mu\,,\,B_{\mu\nu}\,j^\nu\,)\quad j \in \mathcal{A}_J^{Mxw}$$

with kernel

$$B_{\mu\nu}(x) = W_{\mu\nu}\,(x;\zeta)\,ln\,(\mu^2\,(x^2 + i\,o\,x^o\,))\quad . \tag{3.34}$$

The invariant form $(j,\,Bj)$ does not depend on either μ or ζ.

We shall be properly equipped for a systematic study of both the invariant subspace structure and the B-kernels after introducing (in Sec. 3d below) the differential intertwining operators that relate exceptional representations of different Lorentz type in a given sextet.

3.3 Unitary irreducible representations (UIR's)

The bilinear forms constructed in the previous subsection are hermitian for real dimensions d so that the corresponding representations are pseudounitary. Whenever they are positive definite on $\mathcal{A}_\chi$ or on an appropriate invariant subspace (or factor space) of $\mathcal{A}_\chi$ we can assert that the corresponding positive energy representation of $\widetilde{G}$ is unitary. We shall reproduce here without proof the complete list of positive energy UIR's given by Mack [M2], spelling out their Poincaré content and their place in the above classification of elementary induced representations. (For a classification of all UIR's of G see [A1].)

(1) The trivial 1-dimensional representation spans the smallest invariant subspace of $\mathcal{A}_{\chi_o}$, where $\chi_o \Leftrightarrow (d = 0 = j_1 = j_2)$.

(2) The 0-mass particles of helicity 0 or $\pm\frac{1}{2}$ belong to doublet integer points whose Casimir invariants are not identical with any of the finite dimensional ones. Their states span the invariant subspaces $\mathcal{A}_o$ and $\mathcal{A}_{\pm\frac{1}{2}}$ of positive energy solutions of d'Alembert equation (3.16) and of the Weyl equations
$\tilde{\nabla}_{\dot{A}A}\, \phi^A(x) = 0$ for the left handed (neutrino) state (helicity $= j_2 - j_1 = -\frac{1}{2}$);
$\underset{\sim}{\nabla}^{\dot{A}A}\, \varepsilon^*_{\dot{A}\dot{B}}\, \psi^{\dot{B}}(x) = 0\ (= \nabla_\mu\, \psi^{\dot{A}}\, \tilde{\sigma}^\mu_{\dot{A}A_\mu})$ for the right handed state ($j_2 - j_1 = 1/2$) respectively. The (positive invariant) inner product in each of these spaces is given by

$$(\varphi\,,\varphi)_o = i \int_{t=t_o} \overline{\varphi}(t\,,\underline{x})\, \overleftrightarrow{\nabla}_t\, \varphi(t\,,\underline{x})\, d^3x \quad (\varphi \in \mathcal{A}_O)$$

$$(\phi\,,\phi)_{\pm\frac{1}{2}} = \int_{t=t_o} \phi^*(t\,,\underline{x})\, \phi(t\,,\underline{x})\, d^3x \quad (\phi \in \mathcal{A}_{\pm\frac{1}{2}}) \ .$$

The time independence of these expressions is consequence of the conservation of the corresponding current ($i\,\overline{\varphi}\,\overleftrightarrow{D}_\mu\, \varphi$ in the scalar case, $\phi^*\, \tilde{\sigma}_\mu\, \phi$ in the helicity $-\frac{1}{2}$ case, $\psi\, \tilde{\sigma}_\mu\, \psi^*$ in the helicity $\frac{1}{2}$ case).

(3) The 0-mass states of helicity $j_2 - j_1$ exceeding in absolute value 1 belong to invariant subspaces of exceptional representations with $\nu = 1$, $l + 1 = \mid j_2 - j_1 \mid = j_1 + j_2$, $n = 1$ or $2l + 1$ (they belong to the two points at the bottom of the sextet of Fig.1). In other words, we have $j_1 j_2 = 0$, $j_1 + j_2 \geq 1$, $d = j_1 + j_2 + 1$ for this series of representations. A prominent example of this type provided by the dual and the antiselfdual parts of the Maxwell tensor obeying the free Maxwell equations; we have $j_1 + j_2 = 1$, $d = 2$ $(l = 0\,, n = 1)$ in that case; the inner product is expressed in terms of the free 2-point Wightman function of the Maxwell field.

<u>Remark.</u> The representations of these three classes are the only UIR's of the conformal group which remain irreducible when restricted to its Poincaré sugbroup (see [M4] where these representations were termed <u>ladder</u> UIR's). The non-zero helicity (0-mass) representations belong to the holomorphic discrete series of G; they are integrable on G. The latter is not true for the 0-helicity representation.

(4) The minimal subrepresentations of the type of conserved currents with $j_1\,, j_2 > 0$, $d = j_1 + j_2 + 2$ form an invariant subspace of the exceptional integer point $[l + 1\,; l + \nu\,, l + 1 - n]$ (see Fig.1) with $\nu = 1\,, l + 1 = j_1 + j_2$ and n unrestricted ($n = 1, ..., 2l + 1$). The simplest examples of this type are given by the conserved current ($l = 0\,, n = 1$), the stress-energy tensor ($l = 1\,, n = 2$) and the spin $\frac{3}{2}$ counterpart of the stress energy tensor coupled to the gravitino in a supergravity theory ($l = 1/2\,, n = 1$ *and* $n = 2$) (see e.g. [F3]).
When restricted to the Poincaré subgroup such representations exhibit a continuos mass spectrum $0 < m^2 (= -p^2) < \infty$ and a fixed spin $s = j_1 + j_2 (= l + 1)$. Eq. (3.34) provides the positive inner product for the simplest example of this type; its positivity is implied by the relation

$$\int dx \int dy\, \bar{j}^{\mu}(x)\, B_{\mu\nu}(x-y)\, j^{\nu}(y) =$$
$$\int dx \int dy\, \bar{\alpha}^{\mu}(x)\, W_{\bar{V}}(x-y)_{\mu\nu}\, \alpha^{\nu}(y)\ ;\ j^{\mu} = \nabla^{\mu}\nabla\alpha - \Box\alpha^{\mu}\ ,\ \alpha \in \mathcal{A}_V\ .$$

(5) The representations with $j_1 j_2 = 0$ $d > j_1 + j_2 + 1$ contain a continuous mass spectrum, $m^2 > 0$, and spin $s = j_1 + j_2$.

(6) The representations with $j_1 j_2 > 0$, $d > j_1 + j_2 + 2$ involve a continuous mass spectrum $m^2 > 0$, and $s = |j_1 - j_2|, |j_1 - j_2| + 1, ..., j_1 + j_2$.

The last two families are represented in our set in Fig.1 as acting in the (smallest) invariant subspaces of the representation $C_{l+\nu+3}$ for each l, ν, n in the range (3.7).

3.4 Differential intertwining maps between exceptional representations with different Lorentz structure

There are more intertwining operators among exceptional integer points - acting along vertical lines of the sextet diagram of Fig.1. These are the operators $d^{\nu}, d'^{\nu}, \partial_-^n, \partial_+^{2l+2-n}$ in the notation of Fig.1 where

$$d^{\nu} = \left(\frac{1}{\sqrt{2}}\kappa \underset{\sim}{\nabla} \bar{\kappa}\right)^{\nu}\ ,\ d'^{\nu} = \left(\frac{1}{\sqrt{2}}\frac{\partial}{\partial\bar{\kappa}}\tilde{\nabla}\frac{\partial}{\partial\kappa}\right)^{\nu} \tag{3.35a}$$

$$\partial_+ = -\frac{i}{\sqrt{2}}\frac{\partial}{\partial\kappa}\varepsilon \underset{\sim}{\nabla}\bar{\kappa}\ ,\ \partial_- = \frac{i}{\sqrt{2}}\kappa\underset{\sim}{\nabla}\varepsilon^{*}\frac{\partial}{\partial\bar{\kappa}}\ . \tag{3.35b}$$

These maps were introduced and studied in the context of the Euclidean conformal group Spin (5.1) [D3, 4]. We shall not reproduce here the general proof of their invariance but will rather display their properties for simple examples of physical interest.

For tensor type representations d^{ν} amounts to applying (ν times) gradient in x with subsequent symmetrization and subtraction of traces. Its kernel is a finite dimensional invariant subspace of $\mathcal{A}_{\chi^-}$ where

$$\chi^- = \left(d = -l - \nu + 1\,;\ j_1 = l + \frac{1-n}{2}\,,\ j_2 = \frac{n-1}{2}\,;\ \delta = n - l - \nu\right) \tag{3.36}$$

(and all finite dimensional representations of G are obtained in this fashion). d' is, in tensor notation, just a divergence, its kernel being the subspace of conserved tensors.

The sextet displayed on Fig.1 is self-conjugate (i.e. mapped into itself) under space reflection iff $n = l + 1$ and then the operators ∂_+ and ∂_- appear with the same power (n). In this case the direct sum of the inducing representations corresponding to the two bottom points of the sextet diagram gives rise to an irreducible representation of the full Lorentz group (including space reflections). In the case of the Maxwell field (corresponding to $l = 0$, $n = \nu = 1$) this direct sum is spanned by antisymmetric tensors $f_{\mu\nu}$ and

$$\partial_- + \partial_+\ :\ \alpha_{\mu} \rightarrow f_{\mu\nu} = \nabla_{\mu}\alpha_{\nu} - \nabla_{\nu}\alpha_{\mu}\ .$$

With the differential intertwining operators in hand we can see additional structure emerging in the space $\mathcal{A}_V$ (cf. (3.31)). Thus the space $\mathcal{A}_V^{free}$ (3.29) is made up of the invariant subspaces

$$\mathcal{A}_V^{\pm free} = Ker\,\partial_\pm \cap \mathcal{A}_V \qquad (3.37)$$
$$\mathcal{A}_V^{free} = \mathcal{A}_V^{+\,free} \cup \mathcal{A}_V^{-\,free} \;,\; \mathcal{A}_V^{+\,free} \cap \mathcal{A}_V^{-\,free} = \mathcal{A}_V^{long} \;.$$

4 Physical interpretation and possible applications

4.1 General remarks

The mathematical concepts and results of the previous sections have a straightforward interpretation in the context of quantum field theory. It can be sketched as follows.

The vectors in the representation space $\mathcal{A}_\chi$ can be identified with a class of (say, elementary) states in the QFT framework. (For arbitrary χ's they need not correspond to 1-particle states of 0-mass fixed helicity). In general, they are not physical states, but may have the interpretation of states in an indefinite metric space like the one used in any local covariant gauge of QED. If that is the case then we should demand that the representation T_χ is reducible in $\mathcal{A}_\chi$ and contains at least one unitary subrepresentations or factor representation that would single out the physical state space. Thus, we restrict our attention in the applications to labels $(d;\, j_1, j_2)$ that either belong to the list of Sec. 3.3 or are at least partially equivalent to representations of that list.

In order to simplify the language we shall speak in this general introduction about unitary representations leaving aside the discussion of indefinite metric.

Let $\mathcal{H}_1$ be the Hilbert space closure of the space of elementary states. It carries a representation T if $\widetilde{G}$ which is, in general, a direct sum of several (inequivalent) elementary representations T_{χ_i} corresponding to "particles" of different type (including antiparticles). The full QFT space of states $\mathcal{H}(\supset \mathcal{H}_1)$ carries a highly reducible unitary representation $U(g)$ of $\widetilde{G}$ (the "exponential" or "second quantization" of T).

A Poincaré and dilatation covariant local quantum field ψ of Lorentz type $(j_1\,,\, j_2)$ and scale dimension d transforms according to the simple law

$$U(g)\,\psi(x)\,U(g)^{-1} = D^{(j_1,j_2)}(\,\Lambda^{-1}\,)\,\psi(L(\Lambda)\,x + a)\,e^{-\alpha d} \qquad (4.1a)$$

$$g = \begin{pmatrix} e^{\frac{\alpha}{2}}\,\Lambda & i\,\underset{\sim}{a}\,\Lambda^{*-1}\,e^{-\frac{\alpha}{2}} \\ 0 & e^{-\frac{\alpha}{2}}\,\Lambda^{*-1} \end{pmatrix} \;,\; \Lambda \in SL(2\,,\mathbb{C})\,,\; a \in M\;,\; L(\Lambda)\,x = \Lambda\,\underset{\sim}{x}\,\Lambda^{*}. \qquad (4.1b)$$

As shown by Schroer and Swieca [S1] (see also [K6]) a law of this type with $e^{-\alpha d}D^{(j_1,j_2)}(\Lambda^{-1})$ replaced by $D_\chi(g^{-1}\,,\,x)$ (in the notation of Eq.(2.7d)) does not work for a conformal quantum field even in the (generalized) free field case in which

$$\psi(x) = \psi_+(x) + \psi_-(x)$$

where $\psi_\pm$ correspond to the creation and annihilation parts of ψ :

$$\psi_-(x) = 2\pi \int \psi^{(-)}(p)\,e^{ipx}\,\theta_+(p)\,d_4p \;\;,\;\; \psi_- \mid 0 >= 0 \;, \qquad (4.2a)$$

$$\psi_+(x) = 2\pi \int \psi^{(+)}(p)\,e^{-ipx}\,\theta_+(p)\,d_4p \;\;,\;\; < 0 \mid \psi_+ = 0 \;. \qquad (4.2b)$$

The reason is that in the latter case ψ_- and ψ_+ have to transform (in general) under inequivalent representations of $\tilde{G}$; in particular,

$$U(\zeta_1)\,\psi_\pm(x)\,U\,(\zeta_1)^{-1} = e^{i\pi\delta_\pm}\,\psi_\pm(x) \tag{4.3a}$$

$$\delta_+ = d + 2j_2\,(mod\,2)\ ,\quad \delta_- = -\,d + 2j_1\,(mod\,2) \tag{4.3b}$$

ζ_1 being the generating element of the central subgroup $\mathbb{Z}$.

For an arbitrary (interacting) local conformal field we should write it as an expansion in δ

$$\psi(x) = \int_{-1}^{1} d\varrho_\psi(\delta)\,\psi(x,\delta) \ \text{ where }\ U(\zeta_1)\,\psi(x,\delta)\,U(\zeta_1)^{-1} = e^{i\pi\delta}\,\psi(x,\delta)\ : \tag{4.4}$$

here ϱ_ψ is a monotonously increasing function (so that $d\,\varrho_\psi$ is a positive measure). It should agree with the energy positivity of states: if we set

$$\psi(x) = \psi_+(x) + \psi_S(x) + \psi_-(x) \tag{4.5}$$

where $\psi_\pm$ are defined as in (4.2) and ψ_S has support in space like momenta then the expansion (4.4) for $\psi_\pm$ should reduce to a single term with $\delta_\pm$ given by (4.3b).

If we would like to keep the conformal invariance manifest when smearing the field ψ we should introduce a family of test function spaces $\widehat{C}_\delta$ which carry the contragradient representation $\widehat{\chi}$ of χ:

$$\widehat{\chi} = (\,2 - c\ ;\ j_1\,,\,j_2\ ;\ -\,\delta\,)\ \text{ for }\ \chi = (\,2 + c\ ;\ j_1\,,\,j_2\ ;\ \delta\,)\ . \tag{4.6}$$

Then the smeared field is defined by

$$\psi(f) = \int_{-1}^{1} d\,\varrho_\psi(\delta) \int d^4x\,\psi\,(x,\delta)\,f(x,\delta) \equiv \int_{-1}^{1} d\,\varrho_\psi(\delta)\,\psi(f\,,\,\delta)\ ,\quad f(x,\delta)\in\widehat{C}_\delta \tag{4.7}$$

and satisfies the conformal covariance law

$$U(g)\,\psi(f)\,U(g)^{-1} = \int_{-1}^{1} d\,\varrho_\psi(\delta)\,\psi\,(T_{\widehat{\chi}}\,f\,,\,\delta)\ . \tag{4.8}$$

The product $\psi\,f$ in the integrand of (4.7) is meant as Lorentz invariant contraction:

$$\psi\,f = \psi^{A_1\ldots A_{2j_1}\,\dot{B}_1\ldots\dot{B}_{2j_2}}\,f_{A_1\ldots A_{2j_1}\,\dot{B}_1\ldots\dot{B}_{2j_2}} \tag{4.9a}$$

where

$$f_A = \varepsilon_{A\,B}\,f^B\ \ etc. \tag{4.9b}$$

Note that in order to be able to exhibit the local properties of ψ we need the entire test function space $\widehat{C}_\delta$ ($= C_{\widehat{\chi}}$) even if it admits an invariant subspace ($\mathcal{A}_{\widehat{\chi}}$) of positive energy vectors. We observe that the Schwartz space $\mathcal{S}$ of fast decreasing smooth functions is a subspace of each C_χ (its elements having trivial asymptotic expansions; note that $\mathcal{S}$ is only invariant under infinitesimal spacial conformal transformations).

The possible presence of infinitely many δ's in the expansion (4.4) is only associated with the ψ_S term in the decomposition (4.5) and does not contribute to the two point Wightman function

$$W_{\psi\,\psi^*}(x-y) = < \psi(x)\,\psi^*(y) >_o = < \psi_-(x)\,\psi^*_-(y) >_o \quad . \tag{4.10}$$

This function gives rise to an intertwining map from $C_{\widetilde{\chi}}$ into $\mathcal{A}_\chi$ where $\chi = (2 + c\,;\ j_1\,,\,j_2\,;\ c+2j_2\,)$, $\widetilde{\chi} = (2-c\,;\ j_2\,,\,j_1\,;\ c+2j_2\,)$: hence it is proportional to the kernel $W_\chi(x\,;\ \kappa^1 ... \overline{\kappa}^2\,)$ of the Knapp-Stein operator (3.10).

If the invariant hermitian inner product defined by $W_{\psi\,\psi^*}$ on $C_{\widetilde{\chi}}/\,Ker\,W_\chi$ is positive we can identify (the Hilbert space closure of) $C_{\widetilde{\chi}}/\,Ker\,W_\chi$ with a space of "elementary" physical states. An alternative realization of the space of elementary states is given by the space $\mathcal{A}_\chi$ where the inner product is defined in terms of $W_{\widetilde{\chi}^+}$ where

$$\widetilde{\chi}^+ = (\,2-c\,;\ j_2\,,\,j_1\,;\ 2j_1 - c\,) \tag{4.11}$$

(cf. the discussion in Sec. 3.2).

4.2 Current like conformal test function spaces

We can select four invariant subspaces of $C_{\widetilde{V}}$ which are dual to the subspaces of $\mathcal{A}_V \subset C_V$ (see (3.31)) in the sense that they carry equivalent representations. The space $C_{\widetilde{V}}$ modulo $Ker\,W_V$ goes to $\mathcal{A}_V^{o\,long}$ by the map W_V (3.22). We have $C_{\widetilde{V}}/\,Ker\,W_V \simeq Ker\,\overline{W}_{\widetilde{V}}/\,C_J$, where

$$C_J = Ker\,W_V \cap Ker\,\overline{W}_V \tag{4.12}$$

and $\overline{W}_V$, the complex conjugate of W_V, is the intertwining operator which maps $C_{\widetilde{V}}$ into the negative energy subrepresentation space $\overline{\mathcal{A}}_V \subset C_V$. The space C_J which is a good candidate for a test function space for the local electromagnetic potential admits a conformal invariant semidefinite inner product

$$< \phi\,,\,\phi >_\eta = \iint dx_1\,dx_2\,\overline{\phi}^\mu(x_1)\,B_{\mu\nu}(x_{12}\,;\ \zeta\,,\,\eta)\,\phi^\nu(x_2)\ ,\ \ x_{12} = x_1 - x_2\ , \tag{4.13a}$$

where $B_{\mu\nu}$ differs by a new type of longitudinal term from the free 2-point Wightman function (3.32)

$$B_{\mu\nu}(x\,;\ \zeta\,,\,\eta) = W_{\mu\nu}(x\,,\,\zeta) + \frac{\eta}{32\pi^2}\nabla_\mu\nabla_\nu\,[\ln\mu^2\,(x_{12}^2 + i\,o\,x_{12}^o)]^2\ ,\ \ \eta > 0\ . \tag{4.13b}$$

It is straighforward exercise to prove that the hermitian for (4.13) is invariant on C_J and independent of the gauge parameter ζ and of the mass scale μ. Its positive semidefiniteness (for $\eta > 0$) is a consequence of its equivalence to a unitary inner product (see [P6]).

The conserved currents (in the classical field theory language) form an invariant subspace of C_J:

$$C_J^{tr} = Ker\,d' = \{\,\phi^\mu \in C_J\,;\ \nabla_\mu\phi^\mu = 0\,\}\ . \tag{4.14}$$

Its orthogonal complement in C_J with respect to the inner product (4.13) is also invariant and nontrivial):

$$C_J^\perp = \{\,f^\mu \in C_J\,;\ < \phi\,,\,f >_\eta = 0 \quad \text{for all} \quad \phi \in C_J^{tr}\,\}\ . \tag{4.15}$$

The subspace $C_J^{\perp} \cap \mathcal{A}_{\tilde{V}} = \mathcal{A}_{\tilde{V}}$ modulo the subspace of Maxwell currents (see (3.26)) carries a reperesentation which is equivalent to the representation in $\mathcal{A}_V^{long} / \mathcal{A}_V^{olong}$ (cf. (3.31)). The subspace $\mathcal{A}_{\tilde{V}}$ supports an invariant form defined by the longitidinal part of (4.13b). We shall denote by C_J^{Phys} the space C_J^{tr} modulo its subspace $C_J^o \cap C_J^{tr} \supset C_J^{Mxw} (\supset \mathcal{A}_J^{Mxw})$ where C_J^o is the subspace of degeneracy of (4.13):

$$< j , \phi >_\eta = 0 \quad \text{for all} \quad \phi \in C_J \ , \ j \in C_J^o \ . \tag{4.16}$$

The space C_J^{Mxw} is defined as in (3.26) with $\alpha \in C_V$. Note that the space C_J^{tr} (4.14) is dual to $\mathcal{A}_V^{CGF} \cup \overline{\mathcal{A}}_V^{CGF}$ (cf. (3.28)) where $\overline{\mathcal{A}}_V^{CGF}$ is an invariant subspace of $\overline{\mathcal{A}}_V \subset C_V$ for the negative energy subrepresentation. The space C_J^{tr} is equivalently defined as

$$C_J^{tr} = Im\, \partial_+ \, U\, Im\, \partial_- = \{ \, j_\mu(x) = \nabla^\nu f_{\mu\nu}(x) \, , \, f \in C_F \, \} \tag{4.17a}$$

where $\partial_\pm$ are the differential operators of (3.35b) acting in $C_{\tilde{\chi}_F}$, C_{χ_F};

$$C_F = C_{\chi_F} \oplus C_{\tilde{\chi}_F} \ , \ \chi_F = \{ \, d = 2 \, ; \, j_1 = 1 \, , j_2 = 0 \, ; \, \delta = 0 \, \} \tag{4.17b}$$

compare with (3.37).

On C_J^{tr} the form (4.13b) reduce to its "transverse" part given by $B_{\mu\nu} \, (x \, ; \, \zeta = 1 \, ; \, \eta = 0)$. A non-degenerate invariant form on $\mathcal{A}_J^{Mxw} \simeq \mathcal{A}_V / \, Ker\, W_{\tilde{V}}$ is defined in (3.34).

The space C_J^{Phys} is isomorphic to the subspace $\mathcal{A}_F^{free} \subset \mathcal{A}_F \subset C_F^{Phys}$ of positive-energy solutions of the free Maxwell equations, which appear as the image $Im\, \partial_+ \cup Im\ \partial_-$ where $\partial_\pm$ are restricted to $\mathcal{A}_V^{free}$ (cf. (3.29)).

The space $C_J^{Phys} \cup (\mathcal{A}_{\tilde{V}} / \mathcal{A}_{\tilde{V}}^{Mxw}) \simeq C_J / \, C_J^o$ will be used to build the Hilbert space $\mathcal{H}_1$ of single photon states. Let the form $\overline{B}_{\mu\nu}$, complex conjugate to (4.13), be degenerate on $\overline{C_J^o}$. Then every $v \in C_J$ can be split into $v = v_+ + v_- + v_o$ where

$$v_+ \in C_J / \, C_J^o \quad , \quad v_- \in C_J / \, \overline{C_J^o} \tag{4.18}$$

and both v_o and $\overline{v}_o$ belong to the subspace C_J^o of degeneracy of (4.13).

4.3 A conformal invariant formulation of free quantum electrodynamics

It is well-known that the theory of a free electromagnetic field admits a local covariant formulation without recourse to indefinite metric in terms of the Maxwell stress tensor $F_{\mu\nu}$. The drawback of such an approach stems from the fact that it does not seem to have an easy extension to interacting fields. It may be amusing to realize that the constructions of the previous subsection allow us to introduce free smeared electromagnetic potentials with test functions from C_J and a Fock type photon state space in such a way that no indefinite metric is encountered even though some longitudinal degrees of freedom are present.

Since the electromagnetic potential $A_\mu(x)$ is assumed to be hermitian and the representation $\chi_v = [\, d = 1 \, ; \, j_1 = j_2 = \frac{1}{2} \, , \, \delta = 0 \, (mod\, 2)]$ is real, it is natural to single out the set of real test functions in C_J. If A_μ is generalized free field then the smeared field

$$A(v) = \int A_\mu \, (x) \, v^\mu \, (x) \, d^4 x \tag{4.19a}$$

only has contributions from the parts $v_\pm$ of v (see (4.18))

$$A(v) = A^{(+)}(v_+) + A^{(-)}(v_-) \tag{4.19b}$$

with $v_-(x) = \overline{v}_+(x)$ for real v's. The restriction of the test function space to $C_J\,(\subset C_{\widetilde{V}})$ ensures the invariance of the hermitian form

$$< 0 \mid A(\overline{v}_1)\, A(v_2) \mid 0 > = < v_1 \, , \, v_2 >_\eta \tag{4.19c}$$

given by the physical Wightman function (4.13b) (As observed in Sec. 3.2 only the trivial longitudinal 2-point function (3.22) gives rise to a conformally invariant hermitian form in the whole space $C_{\widetilde{V}}$.)

We define the space $\mathcal{H}_1$ of single photon states as the Hilbert space completion of the factor space

$$C_J \,/\, C_J^{\circ} \tag{4.20}$$

with respect to the inner product (4.13). The full Hilbert space H of the free quantum field A_μ is the Fock expotential of H_1:

$$H = \exp H_1 = \{\Phi\} = \{\, \Phi^{(0)},\, \Phi^{(1)}(x_1), \ldots, \Phi^{(n)}(x_1, \ldots, x_n), \ldots\} \tag{4.21}$$

where $\Phi^{(0)} \in \mathbb{C}$, $\Phi^{(n)}$ belongs to the symmetrized tensor product $H_n = H_{1\,sym}^{\otimes n}$ of n copies if H_1 with inner product

$$< \Phi \, , \Phi > = \overline{\Phi}^{(0)}\, \Phi^{(0)} + \int\!\!\int \overline{\Phi}^{(1)}(x_1)\, B(x_{11'})\, \Phi^{(1)}(x_1')\, dx_1\, dx_1' + \ldots + \tag{4.22}$$
$$+ \int \ldots \int \overline{\Phi^{(n)}}(x_1, \ldots, x_n)\, B(x_{11'}) \ldots B(x_{nn'})\, \Phi^{(n)}(x_{1'}, \ldots, x_{n'})\, dx_1\, dx_{1'} \ldots dx_n\, dx_{n'}$$

where $\Phi^{(n)}$ are symmetric functions of their arguments, $B(x_{kk'})$ are the functions (4.13b) (and the Lorentz indices as well as the gauge parameters ζ and η are suppressed). The field $A(f)$ is then defined (as an unbounded operator in H) by

$$[\, A(v)\, \Phi\,]^{(n)}(x_1, \ldots x_n) = \sqrt{n+1} \int\!\!\int v_-(x)\, B(x-y)\, \Phi^{(n+1)}(y\, , x_1\, , \ldots x_n)\, dx\, dy +$$
$$+ \frac{1}{\sqrt{n}} \sum_{k=1}^{n} \Phi^{(n-1)}(x_1\, , \ldots, \widehat{x_k}\, , \ldots, x_n)\, v_+(x_k) \tag{4.23}$$

(where the sign $\frown$ over the argument x_k means that it is absent in $\Phi^{(n-1)}$ and $v_\pm$ are given in (4.18)).

The Maxwell field

$$F_{\mu\nu}(x) = \nabla_\mu A_\nu(x) - \nabla_\nu A_\mu(x) \tag{4.24}$$

is defined as a linear functional on antisymmetric tensor test functions $f^{\mu\nu}$ that belong to C_F (cf. (4.16b)). The action of

$$F(f) = \int F_{\mu\nu}(x)\, f^{\mu\nu}(x)\, d^4x$$

on H can be given by using the Maxwell equations (4.16a)

$$F(f)\,\Phi = A(v)\,\Phi \quad \text{with} \quad v = \nabla\, f^{\mu\nu} \;\; (\in C_J^{tr}) \tag{4.25}$$

Similarly, we define the current as an operator valued distribution on C_V

$$J(\alpha) = \int \alpha^{\mu}(x)\, J_{\mu}(x)\, d^4x \;\; , \quad \alpha^{\mu}(x) \in C_V \tag{4.26}$$

by

$$J(\alpha)\,\Phi = F(f)\,\Phi = A(v)\,\Phi \;\; , \;\; f^{\mu\nu} = \nabla^{[\mu\nu]}\,\alpha \;\; , \;\; v^{\mu} = \nabla_{\nu}\, f^{\mu\nu} \;\; (\in C_J^{Mxw}) \;\; . \tag{4.27}$$

We notice that there is no restriction on the test function spaces for the gauge invariant operators $F_{\mu\nu}$ and J_{μ}. In particular, they have the standard meaning of the local fields since C_F and C_V include all smooth test functions of compact support. (Only the test functions v for the gauge dependent 4-potential A_{μ} are restricted to the subspace C_J of $C_{\widetilde{V}}$ (see (4.12)).

The free Maxwell equations are now reduced to the vanishing of the current $J(\alpha)$ in $\mathcal{H}$ which is a consequence of the degeneracy of the hermitian form B (4.13) on C_J^{Mxw}.

Thus the space$\mathcal{H}$ satisfies all Strocchi-Whightman requirements [S8] [B6] [M6] [M7] for the physical Hilbert space (except for the implicit assumption about the space of test functions for A_{μ}). On the other hand it contains some longitudinal photon states (coming from the elements of $\mathcal{A}_{\widetilde{V}} \,/\, \mathcal{A}_J^{Mxw}$ in $\mathcal{H}_1$). It is natural to define the physical 1-particle space as the completion of $C_J^{Phys} \simeq \mathcal{A}_{F_-}^{free}$ the space of transverse (current-like) test functions. Equivalently, the physical subspace can be defined as the completion of the part D_{Phys} of the field domain in which the conformal generalization of the Gupta-Bleuler [G3] [B5] condition holds in the mean:

$$< \Phi\, , \Box\, \nabla\, A(s)\,\Psi > = - < \Phi\, , A_{\mu}\, (\nabla^{\mu}\, \Box\, s)\,\Psi > = 0 \tag{4.28}$$

for $s \in C_{\chi_o}$ $(\chi_o = (0;0,0;0))$. The eq. (4.28) is a consequence of (4.15) since $\nabla^{\mu}\,\Box\, s \in C_J^{\perp}$ [P6].

The above treatment of free QED is not the only possible one within the present framework, since we did not use the entire space $C_{\widetilde{V}}$. A way to incorporate more sextet spaces, without destroying the conformal invariance of the transverse part of the bilinear form (4.13) that involves the free 2-point Wightman function is to use direct sum of spaces. We shall briefly sketch this possibility.

We define the (large, indefinite metric) 1-particle space as the set of triples

$$\Phi(v) = (\, f_{\mu\nu}(x)\, , h(x)\, , \alpha_{\mu}(x)\,) \in \mathcal{A}_F^{free} \oplus \mathcal{A}_{\widetilde{\chi}_o} \oplus \mathcal{A}_V^{o\, long} \tag{4.29a}$$

where

$$f_{\mu\nu}(x) = \nabla_{[\mu} \int B_{\nu]\varrho}(x-y)\, v_1^{\varrho}(y)\, dy \;,\;\; h = \nabla_{\mu}\, v_{1+}^{\mu} \;,\;\; \alpha_{\mu}(x) = \int W_{v\mu\nu}(x-y)\, v_2^{\nu}(y)\, dy$$
$$v = (\, v_1 \, , \, v_2\,) \;\; , \;\; v_1 \in C_J \;\; , \;\; v_2 \in C_{\widetilde{V}} \;\; , \tag{4.29b}$$

W_V and B are given by (3.22) and (4.13b) $\widetilde{\chi}_o = (4;0,0;0)$ and v_+ as defined in (4.18) belongs to the factor space $C_J \,/\, C_J^o$. On each of the subspaces in (4.29a) an invariant form inherited from the invariant forms on C_J^{tr}, $\mathcal{A}_{\widetilde{V}}$ and $C_{\widetilde{V}}$ can be defined.

Note that the analog of the gauge condition (4.28) holds in the entire subspace built from $\mathcal{A}_F^{free} \oplus \mathcal{A}_V^{o\, long}$ (which contains non physical negative norm vectors) and hence cannot serve for singling out the physical subspace (in contrast to the previous case).

4.4 Prospectives. Discussion of recent work

The representation theoretic part of these lectures is providing an instrument for working in massless QED and in other conformal (gauge) field theories. The discussion of the toy example of the free electromagnetic field in the previous subsection may just serve as simple illustration of what is involved. The question is whether there is a nontrivial conformal QED and if so how it would look like. The second problem seems easier to begin with, but even here the picture is far from being clear. Several approaches have been tested and we shall briefly review some of the most recent ones without committing ourselves as to what is the best starting point.

It is characteristic for new attempts to work with field equations in Minkowski space that they go beyond the elementary repersentations considered here. In the work of Sotkov and Stoyanov [S5-7] nonlinear realizations of the conformal group are used. They propose, in particular, a generalization of the gauge condition (3.28) valid in the interacting case (see also [M5]). It turns out that such nonlinear realizations can be induced by non-decomposable (reducible) reperesentations of our inducing subgroup H and hence appear as a generalization of the EIRs. Similar results were obtained lately by Binegar et al. [B4] starting from a manifestly covariant formulation of the theory (using homogenous functions on the light cone in 6 dimensions - see [M3]), that also involves non-decomposable reperesentations of G. A systematic study of a manifestly conformally covariant action principle is currently under way ([F9]).

The problem of constructing a covariant operator equation in an indefinite metric space which reduces to the Maxwell equations in the physical subspace can also be stated in terms of the (Schwinger) Green functions. It has been addressed in the Euclidean framework in [K7] [F6]. A construction reminiscent to the direct sums of inner product spaces, discussed in the previous subsection,has been proposed in [K7] [P2]. It is based on the study of the exceptional EIRs of the Euclidean conformal group (see [D2-4] as well as remark on p.231 of [T1] where their relevance to the formulation of conformal QED has been first pointed out). Non-decomposable representations of the de Sitter group SO(3,2) have been used to study gauge field models in [F5], [A2], [B3].

Finally, outside the QED problem, we mention that the sextets of exceptional representations (or rather their supersymmetric generalizations) appear to be an ingredient in a systematic approach (followed by G. Sotkov) to the construction of auxiliary fields in supergravity.

References

[A1] E.Angelopoulos, On the unitary dual of $\overline{SO_\nu}(2r, 2)$, Lett.Math. Phys. **7**, 121 - 127 (1983); see also Comm. Math. Phys. (to appear).

[A2] E.Angelopoulos, M.Flato, C.Fronsdal, D.Sternheimer, Massless particles, conformal group and de Sitter Universe, Phys. Rev **D23**, 1278 - 1289 (1981).

[B1] H. Bateman, The transformation of the electrodynamical equations, Proc. London Math. Soc. **8**, 223-264 (1910).

[B2] F. Bayen. M. Flato. Remarks on conformal space, J. Math. Phys. **17**, 1112-1114 (1976).

[b3] B. Binegar, M. Flato, C. Fronsdal, S. Salamo, de Sitter and conformal field theories, Czech. J. Phys. **B32**, 439 - 471 (1982)

[B4] B. Binegar, C. Fronsdal, W. Heidenreich, Conformal QED, preprint UCLA/82/Tep/6 (1982).

[B5] K. Bleuler, Eine neue Methode zur Behandlung der longitudinalen und scalaren Photonen, Helv. Phys. Acta **23**, 567 - 586 (1950).

[B6] P. J. Bongaarts, Maxwell's equations in axiomatic quantum field theory. I. Field tensor and potentials, J.Math. Phys. **18**, 1510 - 1516 (1977); II. Covariant and noncovariant gauges, ibid. **23**, 1181 - 1198 (1982)

[B7] L. Bonora, G. Sartori and M. Tonin, Conformal covariant operator product expansion, Nuovo Cimento **10A**667 - 681 (1972).

[B8] P. Budinich, On conformal spinors and semi spinors, ISAS Preprint 14/82/E.P., Trieste 1982.

[B9] P. Budinich, P. Furlan, C_{2n} spinor geometry, ISAS Preprint 14/82/E.P., Trieste 1982.

[B10] P. Budinich, P. Furlan, On Dirac-like equations in $2n$-dimensional space I, Nuovo Cimento **70A**, 243 - 272 (1982; II ISAS preprint 49/82/E.P., Trieste 1982.

[C1] E. Cartan, Leçons sur la théorie des spineurs (Paris, Hermann, 1937) (English transl.: The Theory of Spinors; with a foreword by R. Streater (Paris, Hermann, 1966) 157 p.)

[C2] N. S. Craigie, V. K. Dobrev, I. T. Todorov, Conformal techniques for OPE in asymptotically free quantum field theory, preprint IC/82/63, Trieste (1982); Conformal covariant composite operators in quantum chromodynamics, Preprint IC/83/35, Trieste (1983).

[C3] E. Cunningham, The principle of relativity in electrodynamics and an extension thereof, Proc. London Math. Soc. **8**, 77-98 (1910).

[D1] P. A. M. Dirac, Wave equations in conformal space, Ann. Math. **37**, 429 - 442 (1936).

[D2] V. K. Dobrev, V. B. Petkova, S. G. Petrova, I. T. Todorov, Dynamical derivation of vacuum operator-product expansion in Euclidean conformal quantum field theory, Phys. Rev. **D13**, 887 - 912 (1976).

[D3] V. K. Dobrev, G. Mack, V. B. Petkova, S. G. Petrova, I. T. Todorov, **Harmonic Analysis on the n -Dimensional Lorentz Group and its Application to Conformal Quantum Field Theory**, Lecture Notes in Physics **63** (Springer, Berlin 1977).

[D4] V. K. Dobrev, V. B. Petkova, Elementary representations and intertwining operators for the group $SU^*(4)$, Rep. Math. Phys. **13**, 233 - 277 (1978).

[F1] S. Ferrara, R. Gatto, A. Grillo, **Conformal Algebra in Space-Time and Operator Product Expansion**, Springer Tracts in Modern Physics, vol. **67** (Springer , Berlin 1973).

[F2] S. Ferrara, R. Gatto, A. Grillo, G. Parisi, General consequences of conformal algebra, in: **Scale and Conformal Symmetry in Hadron Physics**, ed. by R. Gatto (Wiley, N. Y. 1973) pp 59 - 108.

[F3] S. Ferrara, B. Zumino, Structure of linearized supergravity and conformal supergravity, Nucl. Phys. **B134**, 301 - 326 (1978).

[F4] M. Flato, D. Sternheimer, Rémarques sur les automorphismes causals de l'espace temps, Comptes Rendus Acad. Sci., Paris **263a**, 935 - 936 (1966).

[F5] M. Flato, C. Fronsdal, Quantum field theory of singletons, The Rac, J. Math. Phys. **22**, 1100 - 1105 (1981).

[F6] E. S. Fradkin, A. A. Kozhevnikov, M. Ya. Palchik, A. A. Pomeransky, Maxwell equations in conformal invariant electrodynamics, Institute of Electrometry and Automation, preprint n 156, Novosibirsk (1982).

[F7] E. S. Fradkin, M. Ya. Palchik, Recent developments in conformal invariant QFT, Phys. Reports **44C**, 249 - 349 (1978).

[F8] P. Furlan, Are $SO_{2N,2}$ - covariant spinor wave equations also " conformal " invariant ?, Nuov. Cim. **71A**, 43 - 71 (1982).

[F9] P. Furlan, V. B. Petkova, G. M. Sotkov, I. T. Todorov, Conformal quantum electrodynamics with a 5 -potential, ISAS preprint 83/E. P., Trieste (1983).

[G1] A. M. Gavrilik, A. H. Klimyk, Analysis of the representations of the Lorentz and Euclidean groups of n-th order, preprint ITP-75-18E, Kiev (1975).

[G2] P. I. Golod, Conformal group representations in spaces of finite-component fields, preprint ITP-81-133P, Kiev (1981) (in Russian).

[G3] S. N. Gupta, Theory of longitudinal photons in quantum electrodynamics, Proc. Phys. Soc. London **A63** 681 - 691 (1950).

[J1] H. P. Jakobsen, M. Vergne, Wave and Dirac operators and representations of the conformal group, J. Func. Anal. **24** 52 - 106 (1977).

[J2] H. P. Jakobsen, On singular holomorphic representations, Inventiones Math. **62** 67 - 78 (1980).

[K1] A. U. Klimyk, **Matrix Elements and Clebsch -Gordan Coefficients of Group Representations** (Naukova dumka, Kiev 1979, in Russian).

[K2] A.W.Knapp, B.Speth, Status of classification of irreducible unitary representations, in: **Harmonic Analysis**, Lecture Notes in Mathematics **808**, (Springer, Berlin 1982) pp. 1 - 38.

[K3] A.W.Knapp, B.Speh, Irreducible unitary representations of SU(2,2), J. Funct. Anal. **45**, 41 - 73 (1982).

[K4] A.W.Knapp, E.M.Stein, Intertwining operators for semi-simple groups, Ann-Math. **93**, 489-578 (1971).

[K5] A.W.Knapp, G. Zuckermann, Classification theorems for representations of semi-simple Lie groups, in: **Non-commutative Harmonic Analysis**, Lecture Notes in Mathematics **587** (Springer, Berlin 1977) pp. 138 - 159.

[K6] B. Konopelchenko, M.Palchik, Conformal invariance and anomalous dimensions, Sov.J.Nucl.Phys. (transl.) **19**, 106 - 111 (1974) and references therein.

[K7] A.A.Kozhevnikov, M.Ya.Palchik, A.A.Pomeransky, Institute of Electrometry and Automation preprint n. 146, Novosibirsk (1981).

[L1] M.Lüscher, G. Mack, Global conformal invariance in quantum field theory, Comm.Math.Phys. **41**, 203-234 (1975).

[M1] G. Mack, Group theoretical approach to conformal invariant quantum field theory, in **Renormalization and Invariance in Quantum Field Theory**, ed. Caianello (Plenum Press, New York 1974).

[M2] G,Mack, All unitary representations of the conformal group SU(2,2) with positive energy, Comm.Math.Phys. **55**, 1-28 (1977).

[M3] G.Mack, A. Salam, Finite component field representations of the conformal group, Ann.Phys. (N.Y.) **53**, 174-202 (1969).

[M4] G.Mack, I.T.Todorov, Irreducibility of the ladder representations of U(2,2) when restricted to the Poincaré subgroup, J.Math.Phys. **10**, 2078-2085 (1969).

[M5] D.H.Mayer, Vector and tensor fields of conformal space, J. Math. Phys. **16** 884 - 893 (1975).

[M6] M.Mintchev, Quantization in indefinite metric, J.Phys. A: Math.Gen. **13** 1841-1859 (1980).

[M7] M.C.Mintchev, E. d'Emilio, On the reconstruction of the physical Hilbert spaces for quantum field theories with indefinite metric. J.Math.Phys. **22**, 1267-1271 (1981).

[P1] M.Ya.Palchik, On the dynamic nature of global conformal transformations, Phys. Lett, **66B**, 259-261 (1977).

[P2] M.Ya.Palchik, A new approach to the conformal invariance problem in quantum electrodynamics, Institute of Electrometry and Automation preprint n. 180, Novosibirsk (1982).

[P3] S.M.Paneitz, I.E.Segal, Analysis in space time bundles, I. General considerations and the scalar bundle, J. Funct.Anal. **47**, 78-142 (1982), and **49**, 335 -414 (1982).

[P4] R.Penrose, Conformal treatment of infinity, in: **Relativity, Groups and Topology**, Eds. C.M. de Witt and B. de Witt, les Houches Summer School, 1963 (Gordon and Breach, N.Y. 1964) pp 565-584.

[P5] R.Penrose, Twistor algebra, J.Math.Phys. **8**, 345-366 (1967).

[P6] V.B.Petkova, G.M.Sotkov, Exceptional representations of the conformal group and applications - I, II (Sofia 1982) to appear in Bulg.J.Phys;
V.B.Petkova, G.M.Sotkov, The six-point families of exceptional representations of the conformal group, ISAS preprint 15/83/E.P. Trieste (1983).

[P7] A.M.Polyakov, Non-Hamiltonian approach in the conformal invariant quantum field theory, Transl. JETP **39**, 10-18 (1974).

[R1] J.Rawnsley, W. Schmid, J.A.Wolf, Singular unitary representations and indefinite harmonic theory, preprint (November 1981).

[R2] W.Rühl, Distributions in Minkowski space and their connection with analytic representations of the conformal group, CMP **27**, 53-86 (1972).

[R3] W.Rühl, B.C.Yunn, The transformation behaviour of fields in conformally covariant quantum field theory, Fortschr. d. Physik, **25**, 83-99 (1977).

[S1] B.Schroer, J.A.Swieca, Conformal transformations for quantized fields, Phys.Rev. **D10**, 480-485 (1974).

[S2] I.E.Segal, Causally oriented manifolds and groups, Bull.Am.Math.Soc. **77**, 958-959 (1971).

[S3] I.E.Segal, H.P.Jakobsen, B.Ørsted, S.M.Paneitz, B.Speh, Covariant chronogeometry and extreme distances: Elementary Particles, Proc.Nat.Acad.Sci. USA **78**, 5261-5265.

[S4] B.Speh, Degenerate series representations of the universal group of SU(2,2), J.Funct.Anal. **33**, 95-118 (1979).

[S5] G.Sotkov, D.Stoyanov, On the conformal invariance in quantum electrodynamics, J.Phys. **A13**, 2807-2816 (1980).

[S6] D.Stoyanov, G.Sotkov, A conformally invariant gauge fixing condition in quantum electrodynamics, JINR Report, P2-81-665 (1981) (In Russian).

[S7] G.Sotkov, D.Stoyanov, Conformal quantization of electrodynamics (Sofia, 1982) submitted to J.Phys. A.

[S8] F.Strocci, A.S.Wightman, Proof of the charge superselection rule in local relativistic quantum field theory, J.Math.Phys. **15**, 2198-2224 (1974), Erratum, ibid, **17**, 1930-1931 (1976).

[T1] I.T.Todorov, M.C.Mintchev, V.B.Petkova, **Conformal Invariance in Quantum Field Theory** (Scuola Normale Superiore, Pisa 1978)

[T2] I.T.Todorov, Conformal invariance in (gauge) quantum field theory, in: **Mathematical Problems in Theoretical Physics**, Proc. of the Sixth Intern. Conference in Math. Physics. Berlin (West) Aug. 11-20, 1981, Ed. R.Schrader et al, Lecture Notes in Physics, **153** (Springer,Berlin 1982)

[T3] I.T.Todorov, Conformal description of spinning particles, ISAS preprint 1/81/E.P. Trieste (1981).

[U1] A. Uhlmann, The closure of Minkowski space, Acta Phys. Pol. **24**, 295-296 (1963).

[W1] G. Warner, **Harmonic Analysis on Semi-simple Lie groups, I** (Springer, Berlin, 1972).

[Y1] T. Yao, Unitary irreducible representations of SU(2,2), I and II, J.Math.Phys. **8**, 1931-1954 (1967) and **9**, 1615-1626. (1968).

Conformal Composite Fields and Operator Product Expansions

by

N. Karchev (Sofia)*

and

I.T. Todorov**

Zentrum für interdisziplinäre Forschung der Universität Bielefeld

FRG

* N. Karchev
Department of Physics
University of Sofia
Sofia 1126
Bulgaria

** I.T. Todorov
Institute for Nuclear Research and Nuclear Energy
Bulgarian Academy of Science
Sofia 1184
Bulgaria
(permanent address)

INTRODUCTION

The excitement with conformal quantum field theory (QFT) in the early seventies*) had two sources: a phenomenological one - the nearly canonical scaling law observed in deep inelastic electron-proton scattering, and a theoretical one - the possibility of having a non-trivial conformal QFT with anomalous dimensions corresponding to a simple zero $g = g_{cr}$ of the Callan-Symanzik β-function. Asymptotically free QFT like quantum chromodynamics (QCD) and the $\varphi^*\varphi\,\chi$ model in six dimension (φ^3_6 for short) went against this theoretical hope, since away from the point $g=g_{cr}=0$ (for which the theory is free) the Green functions involve logarithmic terms already at the one loop level. It was not until the 1978 work of Efremov and Radyushkin [4] that the relevance of composite conformal operators constructed in refs.[5],[2] out of free constituent fields, was recognized - in the context of studying the asymptotic behavior of the pion fromfactor in QCD. Only gradually did this work gain the deserved popularity - see [6,7].

The present lecture reviews recent work by N. Craigie, V.Dobrev and one of the authors.[8] It demonstrates that the conformal covariance which implies orthogonality of basic (composite) symmetric tensor fields $O_\ell(x)$ also yields their multiplicative renormalizability (without mixing) at the one-loop level. In other words the fields O_ℓ diagonalize the anomalous dimension matrix.[4,7] The possiblity of interpreting such a diagonalization property in terms of conformal Ward identities[9] was earlier discussed in ref. [10].

---

*) It has been crowded a topic as any in high energy (elementary particle) theory. We shall just refer to the reviews [1,2,3] where relevant earlier work by G. Mack, A.A. Migdal, A.M. Polyakov, M.Ya. Palchik the Italian and the Bulgarian collaborations (among others) is cited.

One can also write down an orthogonal set of higher twist operators (built out of a free massless Dirac field) that do not mix under renormalization at the one-loop level. We argue, however, that the contribution of such composite fields to meson wave functions and to physical light-cone operator product expansions should be supressed, since they violate both chiral and conformal invariance.

For preliminaries about the conformal group and its local elementary representations the reader may consult Secs. 1 and 2 of ref. [11].

1. Composite conformal tensor fields

1A. Conserved tensor currents as bilinear combinations of free 0-mass fields

We consider a set of local composite tensor fields built out of scalar spinor or vector (massless) constituents.

In order to incorporate the φ_6^3-model and to prepare the ground for dimensional regularization we consider a (complex) scalar field φ in 2h space-time dimensions :

$$(1.1)\qquad O_\ell^\varphi(x,z) = O^\varphi(x)_{\mu_1\ldots\mu_\ell} z^{\mu_1}\ldots z^{\mu_\ell} = :\varphi^*(x) D_\ell^{h-1}(z\overleftarrow{\nabla}, z\overrightarrow{\nabla})\varphi(x):$$

$$\ell = 0,1,\ldots .$$

The extension of a spinor or a gauge field to arbitrary dimensions although possible, is technically cumbersome and we shall only write the corresponding expressions for 2h = 4:

$$(1.2a) \qquad O^{\psi}_{\ell}(x,z) = :\bar{\psi}(x)\, \mathcal{D}^2_{\ell-1}(z\overleftarrow{\nabla}, z\overrightarrow{\nabla})\Gamma z\psi(x):, \quad \Gamma^{\mu} = i\gamma^{\mu}, i\gamma^{\mu}\gamma_5, \text{ etc.}$$

$$\ell = 1,2,\ldots$$

$$(1.2b) \qquad O^{G}_{\ell}(x,z) = \sum_a z^{\mu}: G^{a\sigma}_{\mu}(x)\mathcal{D}^3_{\ell-2}(z\overleftarrow{\nabla}, z\overrightarrow{\nabla})G^{a}_{\sigma\nu}(x):z^{\nu},$$

$$\ell = 2,3,\ldots .$$

Here z is a light-like vector ($z^2 \equiv \underline{z}^2 - z_o^2 = 0$), $\nabla_{\mu} = \frac{\partial}{\partial x^{\mu}}$ ($\overrightarrow{\nabla}$ acts to the right, $\overleftarrow{\nabla}$ acts to the left); $\mathcal{D}^{\nu}_n(\alpha,\beta)$ is a homogeneous polynomial of degree n:

$$(1.3a) \qquad (\alpha \frac{\partial}{\partial\alpha} + \beta\frac{\partial}{\partial\beta} - n)\, \mathcal{D}^{\nu}_n(\alpha,\beta) = 0 \ ;$$

we can set

$$(1.3b) \qquad \mathcal{D}^{\nu}_n\, (\alpha,\beta) = (\alpha+\beta)^n\, P_n\, (\frac{\beta - \alpha}{\alpha + \beta}) \quad \text{with} \quad \frac{d^{n+1}}{d\xi^{n+1}}\, P_n(\xi) = 0$$

the upper index ν(= 2,3, h-1) stands for the sum of the field dimension d and its spin*) s (d_{ℓ} = h-1, $d_{\psi} = \frac{3}{2}$, $d_G = 2$; $s_{\varphi} = 0$, $s_{\psi} = \frac{1}{2}$, $s_G = 1$). The Dirac matrices γ_{μ} satisfy, as usual, $[\gamma_{\mu},\gamma_{\nu}]_+ = 2\eta_{\mu\nu}$($\eta$ = diag -+++), $\gamma_5 = i\gamma_o\gamma_1\gamma_2\gamma_3 = i\gamma^1\gamma^2\gamma^3\gamma^o$; if we take $\Gamma^{\mu} = \frac{i}{2}\, \gamma^{\mu}(1-\gamma_5)$, or, equivalently, use lefthanded spinors $\Psi_L = \frac{1}{2}(1-\gamma_5)\Psi_L$, then we can rewrite (1.2a) in terms of 2-component (Weyl) spinors:

for

$$(1.2c) \qquad i\gamma_{\mu} = \begin{pmatrix} 0 & \sigma_{\mu} \\ \tilde{\sigma}_{\mu} & 0 \end{pmatrix}, \quad \gamma_5 = \begin{pmatrix} -1 & 0 \\ 0 & 1 \end{pmatrix} \quad \tilde{\sigma}_o = \sigma_o = \mathbb{1}_2,\ \tilde{\sigma}_i = -\sigma_i,\ i = 1,2,3$$

$$\psi_L = \binom{\psi}{0} = \frac{1-\gamma_5}{2}\, \Psi_L \ ,$$

*more precisely, absolute value of helicity.

(1.2d) $\quad 0^{\psi}_{\ell}(x,z) = -:\psi^*(x)\tilde{z}\,\mathcal{D}^{h}_{\ell-1}(z\overleftarrow{\nabla}, z\overrightarrow{\nabla})\psi(x):, \quad \tilde{z} = z^{0} - \underline{z\sigma}.$

(We skip consistently the internal indices of ψ, including the color index in the case of QCD.) The index "a" in (1.2b) runs over the number of generators of the gauge group. If $G^{a}_{\mu\nu}$ stands for the gluon stress tensor (i.e. if the gauge group is color SU(3)) Then "a" runs from 1 to 8. (Eq. (1.2b) also applies to QED, then $G^{a}_{\mu\nu}$ is to be replaced by $F_{\mu\nu}$.)

The expressions (1.1)(1.2) are only suitable for free (massless) fields, satisfying

(1.4) $$\Box\psi = 0 = \Gamma\,\nabla\,\psi, \quad \nabla_{\mu}G^{\mu\nu}_{a} = 0 = \nabla^{\nu}G^{\lambda\mu}_{a} + \nabla^{\lambda}G^{\mu\nu}_{a} + \nabla^{\mu}G^{\mu\lambda}_{a} \quad .$$

We notice that each component of ψ and $G^{\mu\nu}_{a}$ also satisfies the d'Alembert equation as a consequence of (1.4). The necessary modification of the composite operators for interacting constituent fields are discussed in Sec. 2 (see also (1.5) below).

The composite of 0_{ℓ} can be determined (up to a constant factor) from either of the following three requirements: (i) 0_{ℓ} transforms under an elementary representation of the conformal group SO(2h,2) (of scale dimension $d_{\ell} = 2h + \ell - 2$ or twist $d_{\ell} - \ell = 2h - 2$); (ii) the symmetric traceless tensor $0_{\ell}^{\mu_1 \dots \mu_{\ell}}(x)$ satisfies the conservation law $\nabla_{\mu_1} 0_{\ell}^{\mu_1 \dots \mu_{\ell}}(x) = 0$; (iii) in the 1-loop approximation (of either φ^3_6 or QCD) $0_{\ell}(x,z)$ does not mix with $(z\nabla)^{\ell-n}\, 0_n(x,z)$ (for $n < \ell$) under renormalization (in the $\varphi^*\varphi\chi$ model 0^{φ}_{ℓ} can only mix for even ℓ with 0^{χ}_{ℓ}, while in QCD 0^{ψ}_{ℓ} mixes with 0^{G}_{ℓ} for $\ell = 2,4,\dots$. We shall elaborate on this latter property in Sec. 2 below. It is the gauge invariant counterparts of 0^{ψ}_{ℓ} and 0^{G}_{ℓ} (obtained from (1.2) by replacing ∇_{μ} by the covariant derivative

(1.5a) $$D_{\mu} = \nabla_{\mu} + A_{\mu}$$

where

(1.5b) $$A_\mu = \sum_{a=1}^{\infty} g \frac{\lambda_a}{2i} A_\mu^a$$

for a spinon field in QCD, and

(1.5c) $$(A_\lambda G_{\mu\nu})^a = gf^{abc} A_\lambda^b G_{\mu\nu}^c$$

when applied to the gluon field) whose renormalization properties will be studied in the case of QCD. The conformal invariance requirement ((i) in the above list) is the most general one, since it does not use the equations of motion. The corresponding derivation is given in Appendix A. The fact that we effectively replace the multi-index tensor formalism by a systematic use of homogeneous polynomials of type (1.1) or (1.2) makes ready at our disposal standard analytic tools [12,13,14]. The traceless tensors are in one-to-one correspondence with homogeneous polynomials of a light-like vector (since the latter admit a unique harmonic continuation - see [13]). In terms of these variables the conservation law assumes the form

(1.6a) $$\nabla\delta\, O_\ell(x,z) = 0$$

where

(1.6b) $$\delta_\mu = (h - 1 + z\partial)\partial_\mu - \frac{1}{2} z_\mu \partial^2 \quad (\partial_\mu = \frac{\partial}{\partial z^\mu}\, ,\ \partial^2 = \frac{\partial}{\partial z_\lambda \partial z^\lambda}$$

is the interior derivative on the light cone[13]. If $f_1(z)$ and $f_2(z)$ are two polynomials that coincide on the cone $z^2 = 0$, so that $f_1(z) - f_2(z) = z^2 f(z)$, then $\delta_\mu(f_1 - f_2)|_{z^2 = 0} = 0$, since

(1.6c) $$\delta_\mu z^2 = z^2(\delta_\mu + 2\partial_\mu).$$

Moreover, if we interpret for a moment z as "momentum", we observe that $2\delta_\mu$ have the properties of the (mathematical) generators of special conformal transformations (of the massless 0-helicity representation of the conformal group SO(2h,2):

(1.6d) $$[\delta_\mu, \delta_\nu] = 0, \quad [\delta_\mu, z_\nu] = \eta_{\mu\nu}(h-1+z\partial)+z_\nu\partial_\mu - z_\mu\partial_\nu ,$$

(1.6e) $$\delta_\mu\delta^\mu = 0.$$

(First order interior differentiations coincide with tangent vector fields on the cone $z^2 = 0$; examples are provided by the generators of dilations and Lorentz transformations in the right-hand side of (1.6d))
Substituting the expression (1.1) for O_ℓ^φ in (1.6) we find the following equation for the homogeneous polynomial $\mathcal{D}_\ell^{h-1}$:

(1.7) $$(p_1+p_2)\delta\, \mathcal{D}_\ell^{h-1}(p_1z,\, p_2z) = 0 = \left(\frac{\partial}{\partial\alpha} + \frac{\partial}{\partial\beta} - \frac{\alpha+\beta}{h+\ell-2}\,\frac{\partial^2}{\partial\alpha\partial\beta}\right)\mathcal{D}_\ell^{h-1}(\alpha,\beta) .$$

In deriving the second relation we have used the free d'Alembert equation for the field φ,

(1.8) $$\Box\varphi(x) = 0 \quad (\Box = \nabla^2)$$

which implies that one only needs to consider light-like p's.

Inserting the representation (1.3b) for $\mathcal{D}_\ell$ in (1.7) we find an ordinary differential equation for P_ℓ:

(1.9) $$(\xi^2-1)\frac{d^2}{d\xi^2} P_\ell(\xi) + 2(h-1)\xi \frac{d}{d\xi} P_\ell(\xi) - \ell(\ell+2h-3)P_\ell = 0$$

(that is special case for $\nu = h-1$ - of Eq. (A.10) of Appendix A).

Its general non-singular solution is proportional to a Gegenbauer polynomial,

(1.10) $$P_\ell(\xi) = N_\ell^{h-1} C_\ell^{h-\frac{3}{2}}(\xi), \text{ so that } \mathcal{D}_\ell^\varphi(\alpha,\beta) = N_\ell^{h-1}(\alpha+\beta)^\ell C_\ell^{h-\frac{3}{2}}\left(\frac{\beta-\alpha}{\beta+\alpha}\right).$$

1B. 2- and 3-point functions of composite conformal fields

The 2-point functions of $O_\ell(x,z)$ are determined (up to a constant factor) from conformal invariance alone (see [2]):

(1.11) $$<O_\ell(x_1,z_1)\, O_n(x_2,z_2)>_o = A_\ell\delta_{\ell n}\int \Theta_+(p)(-p^2)_+^{h-2+\ell}\, \Pi^{\ell\ell}(p;z_1,z_2)\, e^{ipx_{12}}dp.$$

$$(x_{12} = x_1-x_2,\ \Theta_+(p) = \Theta(-p^2),\ px = \mathbf{px} - p^o x^o).$$

Here $\Pi^{\ell\ell}$ is the projection operator on maximal spin (ℓ) in the space of symmetric traceless tensors of rank ℓ:

(1.12) $$p\delta_1,\ \Pi^{\ell\ell}(p;z_1,z_2) = 0 = p\delta_2\Pi^{\ell\ell}(p;z_1,z_2)$$

($\delta_{1,2}$ being the interior derivative with respect to $z_{1,2}$, respectively); in 4-dimensional space time $\Pi^{\ell\ell}$ is proportional to the ℓth Legendre polynomial of the cosine of the angle between $\underline{z}_1$ and $\underline{z}_2$ in the rest frame of p:

$$(1.13)\qquad \Pi^{\ell\ell}(p;z_1,z_2) = \frac{\ell!}{(2\ell-1)!!}\left(\frac{pz_1, pz_2}{-p^2}\right)^{\ell} P_{\ell}\left(1-\frac{p^2 z_1 z_2}{pz_1 pz_2}\right) \quad (\text{for } h = 2)\cdot$$

It satisfies the normalization condition

$$(1.14)\qquad \frac{1}{\ell!}\Pi^{\ell\ell}\left(p;z_1,\frac{\partial}{\partial\zeta}\right)\Pi^{\ell\ell}(p;\zeta,z_2) = \Pi^{\ell\ell}(p;z_1,z_2)$$

(for the corresponding harmonic extension in ζ - see [14], Sec. 5B). The orthogonality property displayed in (1.11) is a consequence of the non-equivalence of the representations of the conformal group carried by O_ℓ and O_n for $\ell \neq n$ (cf. [1,14]). The conservation law (1.12) is related to the canonical dimension

$$(1.15)\qquad d_\ell = 2h - 2 + \ell \quad (\text{or minimal twist } d_\ell - \ell = 2h - 2)$$

of the composite field O_ℓ. We shall see in the next section that one effect of renormalization is the appearance (in general) of an anomalous dimension for the fields O_ℓ, which destroys their conservation (except for the electric current and the stress energy tensor).

A natrual choice for the normalization constant N_ℓ in (1.10) (or (A.11) for the general case) is suggested by the free field operator product expansion. If we set (following [8])

$$(1.16a)\qquad \mathcal{D}^{\nu}_{\ell}(-1,1)\ \left(= N^{\nu}_{\ell}\,\frac{2^{\ell}}{\ell!}\,\frac{d^{\ell}}{d\xi^{\ell}}\,C^{\nu-\frac{1}{2}}_{\ell}(\xi)\right) = 1$$

or

$$(1.16b)\qquad (N_\ell^\nu)^{-1} = \binom{2\nu+2\ell-2}{\ell}$$

then we can write (for $z^2 = 0$)

$$(1.17a)\qquad :\varphi(x + \tfrac{z}{2})\varphi^*(x - \tfrac{z}{2}): = \sum_{\ell=0}^{\infty} \frac{1}{\ell!} R_\ell(x,z)$$

where R_ℓ are "light-ray" composite fields (cf. ref. [15])

$$(1.17b)\qquad R_\ell(x,z) = \frac{(2h + 2\ell - 3)!!}{(2h + 2\ell - 4)!!} \frac{1}{2} \int_{-1}^{1} d\xi \, (1-\xi^2)^{h+\ell-2} \, O_\ell(x+\xi\tfrac{z}{2},z)$$

$$(1.17c)\qquad = O_\ell(x,z) + \frac{1}{2h+2\ell-1} \frac{(z\nabla)^2}{8} O_\ell(x,z) + \ldots .$$

(The passage from the light ray formula (1.17b) to the short distance Wilson expansion (1.17c) can be effected through expanding $O_\ell(x+\xi\frac{z}{2})$ in a power series in ξ. It is only legitimate if O_ℓ is applied on a finite energy vector.) Eq.(1.17) can be verified by using the orthogonality relation (1.11) which implies

$$(1.18a)\qquad i^\ell < :\varphi(x + \tfrac{z}{2})\varphi^*(x - \tfrac{z}{2}): \tilde{O}_\ell(-p,\zeta)>_o = (p\zeta)^\ell \, \hat{C}_\ell^{h-\frac{3}{2}}(2i \frac{\zeta\nabla_z}{\zeta p}) \, I_h(x,z;p)$$

$$(1.18b)\qquad = \frac{i^\ell}{\ell!} < R_\ell(x,z) \, \tilde{O}_\ell(-p, \zeta) >_o ;$$

here $\hat{C}_\ell^{\nu-\frac{1}{2}} \equiv N_\ell^\nu C_\ell^{\nu-\frac{1}{2}}$, $\tilde{O}_\ell(-p) = \int O_\ell(y) \, e^{ipy} dy$,

$$(1.19a)\qquad I_h(x,z;p) \equiv \int <\varphi(x + \tfrac{z}{2})\varphi^*(y)>_o \, <\varphi^*(x - \tfrac{z}{2})\varphi(y)>_o e^{ipy} d^{2h}y$$

(1.19b) $$= \frac{2\pi}{(4\pi)^h}\, \Theta_+(p)\, e^{ipx} \left(-\frac{p^2}{2}\right)^{h-2} f_{h-2}(\omega), \quad \omega = \tfrac{1}{2}\sqrt{(pz)^2 - p^2z^2},$$

and f_n are related to the spherical Bessel functions,

(1.19c) $$f_n(\omega) = \sqrt{\frac{\pi}{2}}\; \omega^{-\frac{1}{2}-n}\, J_{n+\frac{1}{2}}(\omega) = \frac{1}{2(2n)!!} \int_{-1}^{1} \cos\xi\omega(1-\xi)^n\, d\xi =$$

(1.19d) $$= \frac{1}{(2n+1)!!} - \frac{1}{(2n+3)!!}\frac{\omega^2}{2!!} + \frac{1}{(2n+5)!!}\frac{\omega^4}{4!!} - \ldots .$$

2. Renormalization of composite field operators

The change of a composite operator $O_\ell(x,z)$ in the presence of interaction $L_I(x)$ (corresponding to a scattering operator $S = \mathrm{T}\exp\{i\int L_I(x)dx\}$) is given by

(2.1) $$\delta O_\ell(x,z) = T(O_\ell(x,z)S)S^{-1} - O_\ell(x,z)$$

(T stands for time ordered product, as usual). The coefficients in the expansion of the right hand side in terms of normal products of constituent fields are given, in general, by divergent perturbation theoretic integrals, which will be dimensionally regularized. We shall demonstrate that conformal covariance of composite operators persists at the 1-loop level and that they acquire amomalous dimension as a result of the infinite renormalization.
For the sake of simplicity we shall outline the main steps of the calculation for the φ_6^3 model, only indicating afterwards the necessary modifications needed for the treatment of a more realistic (gauge) theory.

2 A. The anomalous dimension matrix for the φ^3_6-model

Taking

(2.2) $$L_I(x) = g: \varphi^*(x)\, \varphi(x)\, \chi(x):$$

and suppressing the variable z (or, equivalently, the tensor indices) we can write (for $\phi = \varphi$ or χ)

(2.3)
$$i^{\ell}\delta O_{\ell}^{\phi}(x) = \int G_{\ell}^{\phi}(x;y)\chi(y)dy +$$
$$+ \int\int \{\Gamma_{\ell}^{\phi\varphi}(x;y_1,y_2): \varphi^*(y_1)\varphi(y_2): + \Gamma_{\ell}^{\phi\chi}(x;y_1,y_2): \chi(y_1)\chi(y_2):\}\, dy_1\, dy_2 + \dots$$

where higher order polynomials in the basic fields are omitted. The first (linear in χ) term has a non-vanishing contribution in the 1-loop approximation for $\ell = 0$ only (due to the orthogonality property of O_ℓ) and has no counterpart for gauge invariant composite operators in QCD. We shall focus our attention on the quadratic term in the fields. The presence of two quadratic structures stems from the fact that local operators with the conformal properties of O_ℓ can be built up (for even ℓ's) not only from φ^* and φ but also from two χ's.

The coefficients Γ_ℓ in the expansion (2.3) coincide with the (amputated) vertex functions. For instance $\Gamma_{\ell}^{\varphi\varphi}$ is given in the 1-loop approximation by the sum of three graphs displayed on Fig. 1 (see p. 11a)

All three graphs are ultraviolet divergent. We shall demonstrate that the coefficient to $\frac{1}{\varepsilon}$ (ε = 3-h) for each of the dimensionally regularized p-space contributions of these diagrams is proportional to the polynomial $\mathcal{D}_{\ell}^{h-1}(zp_1, -zp_2)$ that enters the definition (1.1) of O_{ℓ}^{φ}.

That is obvious for the wave function renormalization graphs of

FIGURE 1

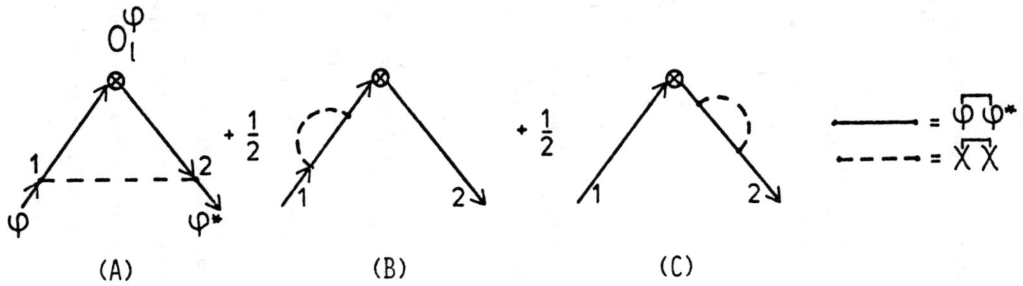

One loop diagrams contributing to the coefficient $\Gamma_\ell^{\varphi\varphi}$ in the expansion of δO_ℓ^φ

Fig. 1B and 1C, whose singular parts are

$$(2.4)\qquad SI_{(B)}(p_1,p_2;z) = SI_{(C)}(p_1p_2;z) = \frac{g^2}{(4\pi)^3}\,\frac{1}{6\varepsilon}\,\mathcal{D}_\ell^{h-1}(p_1z,\,-p_2z).$$

On the other hand the $\frac{1}{\varepsilon}$ contribution of the triangular graph of Fig. 1A can be written in the form

$$(2.5)\qquad SI_{(A)}(p_1,p_2;z) = -\frac{1}{\varepsilon}\frac{g^2}{(4\pi)^3}\int_0^1 d\alpha\int_0^1 \beta\, d\beta\, \mathcal{D}_\ell^{(2)}((1-\beta)zp_1 +$$

$$+\ \alpha\beta z(p_1-p_2),\ -(1-\beta)zp_2+(1-\alpha)\beta z(p_1-p_2))$$

(see Appendix B).

Thus the coefficient to $\frac{1}{\varepsilon}$ is a homogeneous polynomial of degree ℓ, call it $\tilde{\mathcal{D}}_\ell$, of p_1z and $-p_2z$. Inserting (2.5) in the expression for δO_ℓ^φ we find

$$(2.6)\qquad \delta^{(A)}O_\ell^\varphi(x,z) = \frac{1}{\varepsilon}\int\int \varphi^*(x_1)\varphi(x_2)\,\tilde{\mathcal{D}}_\ell(z\nabla_1,z\nabla_2)\delta(x_1-x)\,\delta(x_2-x)dx_1dx_2 +$$

$$+\ \text{finite terms for } \varepsilon\to 0.$$

It is now crucial for our argument*) that the right hand side of (2.6) should be again a conformal covariant (local) field. Indeed, the φ_6^3 -model (with interaction Lagrangian (2.2)) is manifestly conformally invariant. Conformal invariance can only be broken in the process of (perturbative) renormalization when one introduces dimensional parameters (a non-zero subtraction point in p^2) in the finite parts of propagators and vertex functions; there is no such parameter in the $\frac{1}{\varepsilon}$ term (2.5).

---

*) An alternative argument that uses the orthogonality of O_ℓ is presented in ref. [8].

Knowing that $SI_{(A)}$ is proportional to $\mathcal{D}_\ell^{h-1}(zp_1 - zp_2)$ we can evaluate the proportionality coefficient by giving the special value

$$p_1 = p_2 \equiv p \tag{2.7}$$

to the momenta. With the normalization condition (1.16) we find

$$\begin{aligned} SI_{(A)}(p,p;z) &= -\frac{(-pz)^\ell}{\varepsilon}\frac{g^2}{(4\pi)^3}\int_0^1 (1-\beta)^\ell \beta d\beta = \\ &= -\frac{(-pz)^\ell}{\varepsilon}\frac{g^2}{(4\pi)^3}\frac{1}{(\ell+1)(\ell+2)}\ , \end{aligned} \tag{2.8}$$

so that

$$\begin{aligned} SI(p_1,p_2;z) &= SI_{(A)}(p_1,p_2;z) + \frac{1}{2}[SI_{(B)}(p_1,p_2;z) + SI_{(E)}(p_1,p_2;z)] = \\ &= \frac{1}{\varepsilon}\frac{g^2}{(4\pi)^3}\left[\frac{-1}{(\ell+1)(\ell+2)} + \frac{1}{6}\right]\mathcal{D}_\ell^{h-1}(p_1 z, -p_2 z)\ . \end{aligned} \tag{2.9}$$

A similar argument goes through for the other coefficients Γ_ℓ in the expansion (2.3). We find

$$\delta O_\ell^\varphi = \frac{1}{\varepsilon}(\gamma_\ell^{\varphi\varphi} O_\ell^\varphi + \gamma_\ell^{\varphi\chi} O_\ell^\chi) + \dots \tag{2.10a}$$

$$\delta O_\ell^\chi = \frac{1}{\varepsilon}(\gamma_\ell^{\chi\varphi} \theta_\ell^\varphi + \gamma_\ell^{\chi\chi} O_\ell^\chi) + \dots \tag{2.10b}$$

(the dots standing for finite terms for $\varepsilon \to 0$, $\ell > 0$);

$$(2.11) \qquad \gamma_\ell = \begin{pmatrix} \gamma_\ell^{\varphi\varphi} & \gamma_\ell^{\varphi\chi} \\ \gamma_\ell^{\chi\varphi} & \gamma_\ell^{\chi\chi} \end{pmatrix}$$

is the anomalous dimension matrix. Its eigenvalues give the anomalous dimensions of the corresponding linear combinations of O_ℓ^φ and O_ℓ^χ that are renormalized multiplicative by in the 1-loop approximation.

For odd ℓ O_ℓ^χ vanishes and the eigenvalue γ_ℓ^φ corresponding to the operators O_ℓ^φ is read from the right-hand side of (2.9):

$$(2.12) \qquad \gamma_\ell^\varphi \ (= \gamma_\ell^{\varphi\varphi}) = \frac{g^2}{(4\pi)^3} \left[- \frac{1}{(\ell+1)(\ell+2)} + \frac{1}{6} \right] \quad \text{for } \ell = 1,3,5,\ldots.$$

We see, in particular, that the anomalous dimension of the current

$$(2.13) \qquad j_\mu(x) = iT(:\varphi^*(x)\overleftrightarrow{\partial}_\mu \varphi(x):S)S^{-1} = 2T(O_{1\mu}(x)S)S^{-1}$$

vanishes (at least at the 1-loop level) in agreement with current conservation law, while $\gamma_\ell^\varphi > 0$ for $\ell = 3,5,\ldots$ in accord with positivity (see [2]).

We proceed to the diagonalization of the anomalous dimension matrix for even $\ell = 2n$. The graphs contributing to $\Gamma_\ell^{\varphi\chi}$, $\Gamma_\ell^{\chi\chi}$ are pictured on Figs. 2A,2B and 2C respectively (see p.14a) Numerically, all these graphs have the same contributuions as their counterparts on Fig. 1. Therefore, we have

$$(2.14) \qquad \gamma_{2n} = \frac{g^2}{(4\pi)^3} \begin{pmatrix} -\frac{1}{(2n+1)(2n+2)} + \frac{1}{6} & -\frac{1}{(2n+1)(n+1)} \\ -\frac{1}{(2n+1)(2n+2)} & \frac{1}{6} \end{pmatrix}.$$

FIGURE 2

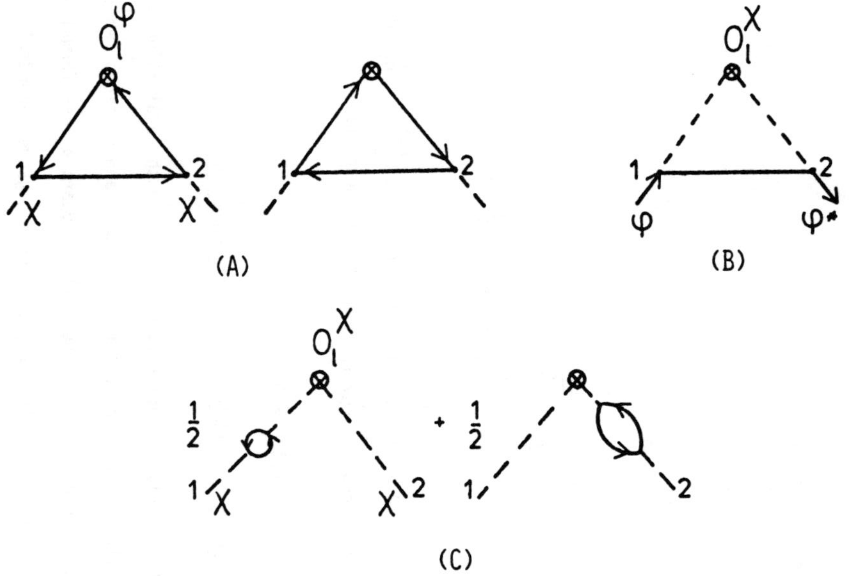

One loop diagrams contributing to (A) $\Gamma_{\ell}^{\varphi X}$, (B) $\Gamma_{\ell}^{X\varphi}$, (C) Γ_{ℓ}^{XX} .

The linear combinations of O^{φ}_{2n} and O^{χ}_{2n} that diagonalize γ_{2n} are

(2.15a) $$O^{(+)}_{2n} = O^{\varphi}_{2n} + \frac{1}{2} O^{\chi}_{2n}$$

(2.15b) $$O^{(-)}_{2n} = O^{\varphi}_{2n} - O^{\chi}_{2n} .$$

We have

(2.16a) $$\delta O^{(\pm)}_{2n} = \gamma^{(\pm)}_{2n} O^{(\pm)}_{2n}$$

where

(2.16b) $$\gamma^{(\pm)}_{2n} = \frac{g^2}{(4\pi)^3} \begin{cases} -\frac{1}{(2n+1)(n+1)} + \frac{1}{6} \\ \frac{1}{(2n+1)2n+2)} + \frac{1}{6} \end{cases} .$$

We observe that $\gamma^{(+)}_{2} = 0$ in accord with the fact that $O^{(+)}_{2}$ coincides with the (conserved) stress energy tensor. We also note that

(2.15c) $$<O^{(+)}_{2n}(x) O^{(-)}_{2n}(y)>_o = 0 .$$

2 B. Remarks concerning the QED and the QCD calculation

Apart from the purely algebraic complication in the evaluation of Feynman graphs with spinor and vector propagators and vertices (which is in effect dealt with in Appendix B) there also appears a less formal one when we proceed to more realistic (gauge) theories: gauge dependent propagators and vertices are not conformal invariant and hence our covariance argument needs a modification. The difficulty comes from the fact that although the gauge invariant, say QED, Lagrangian

(2.17) $$L_{inv} = \frac{1}{2} (\frac{1}{2} F_{\mu\nu} - \nabla_{\mu} A_{\nu} + \nabla_{\nu} A_{\mu}) - \bar{\psi} \not{D} \psi , \quad \not{D} = \not{\nabla} - ie \not{A}$$

is conformally invariant, standard gauge fixing terms (like $-\frac{1}{2}(\nabla A)^2$ for the Gupta-Bleuler gauge) are not.

It is possible to overcome this difficulty by using a conformal family of gauges associated to a non-decomposable representation for the electromagnetic potential (that involves a dimensionless generalized free scalar field)[16,17,18]. For the evaluation of a gauge invariant quantity, however, it is sufficient to know that tree diagrams are conformal invariant up to accompanying (non-local) gauge transformations[2,19], which only contribute to gauge dependent Green functions.

As an illustration we shall evaluate the anomalous dimension matrix for second rank symmetric tensors in QED.

Consider the gauge invariant composite operators

$$(2.18a) \qquad O^{\psi}(z.x) = \frac{1}{2}\, z^{\mu} z^{\nu} \{:\bar{\psi}(x)\gamma_{\mu} \mathcal{D}_{\nu}\psi(x): - :\overline{\mathcal{D}_{\nu}\psi(x)}\ \gamma_{\mu}\psi:\}$$

$$(2.18b) \qquad O^{F}(z,x) = z^{\mu} z^{\nu} :F_{\mu}{}^{\rho} F_{\rho\nu}: \quad .$$

Because of gauge invariance (see Appendix C) the counterpart of Eq. (2.10) in this case would read

$$(2.19) \qquad \delta O^{\psi} = \frac{1}{\varepsilon}\,(\gamma^{\psi\psi} O^{\psi} + \gamma^{\psi F} O^{F})$$

$$\delta O^{F} = \frac{1}{\varepsilon}\,(\gamma^{F\psi} O^{\psi} + \gamma^{FF} O^{F}).$$

The graphs contributing to $\gamma^{\psi F}$ and $\gamma^{F\psi}$ are given on Figure 3 (see p. 16a).

FIGURE 3

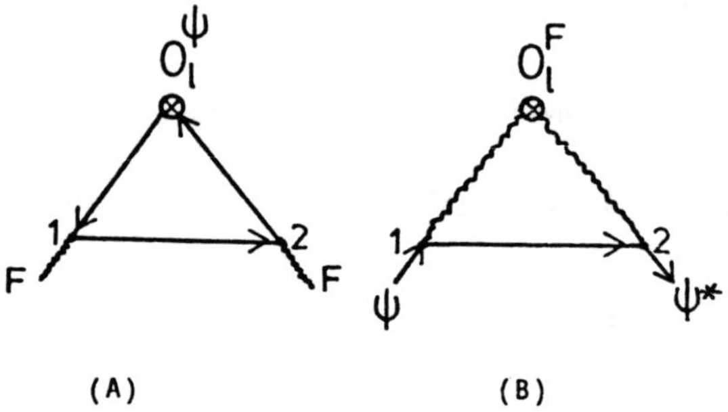

One loop diagrams contributing to (A) $\Gamma_2^{\psi F}$ and (B) $\Gamma_2^{F\psi}$.

At the 1-loop level $\gamma^{FF} = 2\gamma_F$, where γ_F is the anomalous dimension of the photon field. The matrix element $\gamma^{\psi\psi}$ can be represented in the form

$$\gamma^{\psi\psi} = \gamma_A^{\psi\psi} + \gamma_B^{\psi\psi} + 2\gamma_\psi$$

where $\gamma_A^{\psi\psi}$ and $\gamma_B^{\psi\psi}$ are determined by the graphs on Figs. 4A and 4B, and γ_ψ is anomalous dimension of the fermion field.

FIGURE 4

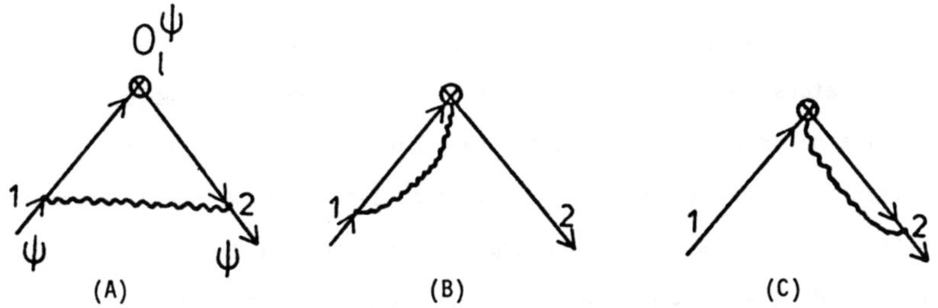

One loop diagrams contributing to $\Gamma_2^{\psi\psi}$.

The gauge invariance of the subtraction term (Appendix C) guarantees the gauge invariance of the anomalous dimension matrix for comuptational convenience the Feyman gauge we obtain

$$\text{(2.20)} \qquad \gamma^{\psi\psi} = \gamma^{F\psi} = \frac{4}{3}\frac{\alpha}{\pi}$$

$$\gamma^{FF} = \gamma^{\psi F} = \frac{2}{3}\frac{\alpha}{\pi}$$

where $\alpha = \frac{e^2}{4\pi}$.

The linear combinations of O^F and O^γ that diagonalize (γ^{AB}) are

$$O^- = O^F - O^\psi \equiv \Theta_{\mu\nu} z^\mu z^\nu$$

$$O^+ = O^F + 2 \cdot O^\psi$$

while $\gamma^- = 0$ and $\gamma^+ = 2\frac{\alpha}{\pi}$. Therefore, the anomalous dimension of the stress energy tensor, that coincides while O^- is zero (at least at 1-loop level) as implied by energy momentum conservation.

3. Application of diagonalized anomalous dimensions to operator product expansion

In the preceding sections we constructed a set of conformal tensor operators. Because they are multiplicatively renormalizable, they form a natural basis for light-cone expansions of products of two currents in the theory of deep inelastic scattering. We shall sketch the main steps involved in such an application. Consider the matrix element of the time-ordered product of two gauge-invariant (for example, electromagnetic) currents between hadron states

$$T_{\mu\nu} = \int d^4z\, e^{-iqz} < p|TJ_\mu(z)J_\nu(0)|p > \tag{3.1}$$

$$= \left(-g_{\mu\nu} + \frac{q_\mu q_\nu}{q^2}\right) T_1 + \frac{1}{m_h^2}\left(p_\mu - q_\mu \frac{p.q}{q^2}\right)\left(p_\nu - q_\nu \frac{p.q}{q^2}\right) T_2$$

where $|p>$ is an on shall sharp momentum hadron state, and a spin average has been carried out. The structure functions of deep inelastic scattering are related by the optical theorem to the imaginary parts of $T_{1;2}$.

The Bjorken limit corresponds to letting $q^2 \to \infty$, $p.q = \nu \to \infty$ with $x = \frac{q^2}{2\nu}$ fixed. This limit is determined by the singularities of the product of two currents on the light cone $z^2 = 0$. The asymptotic behavior of the amplitude (3.1) is exhibited by inserting the light cone expansion

$$(3.2) \qquad TJ^{\mu}(z)J^{\nu}(0) = \sum_{i,n} C^{\mu\nu}_{in}(z^2) z^{\lambda_1} \ldots z^{\lambda_n} O^i_{\lambda_1 \ldots \lambda_n} .$$

The singularities for $z^2 \to 0$ are factored in the C-number coefficients, and O^i are local composite operators. The index i labels fields that carry the same quantum numbers and, therefore, may mix under renormalization. The coefficient functions satisfy the renormalization group equation[20]

$$(3.3) \qquad [\mu \frac{\partial}{\partial\mu} + \beta(g) \frac{\partial}{\partial g}] C^{\mu\nu}_{in}(z^2,g,\mu) = \sum_j \gamma^{(n)}_{ij} C^{\mu\nu}_{jn}(z^2,g,\mu).$$

If the anomalous dimension matrix $\gamma^{(n)}$ is diagonalized - which, as we saw, is greatly facilitated by using conformal composite operators - then Eq. (3.3) can actually be solved in an asymptotically free theory with

$$(3.4) \qquad \beta(g) = -\beta_0 g^3.$$

The result is the free field (Bjorken) value of $C^{\mu\nu}_n$ multiplied by a factor of the order of

$$[\log \frac{1}{\mu^2 z^2}]^{-\frac{\gamma_n}{2\beta_0}}$$

for $z^2 \to 0$ where γ_n is the corresponding anomalous dimension eigenvalue.

ACKNOWLEDGEMENTS

We are grateful to Dr. V. DOBREV for discussions.

One of us (I.T.) would like to thank Professor L. STREIT for his kind invitation to the Zentrum für interdisziplinäre Forschung at the University of Bielefeld where the final version of these notes was completed. A useful discussion with Professor G.F. DELL'ANTONIO at this late stage is also gratefully acknowledged.

Appendix A

Conformal covariant bilinear composite operators

In order to treat the composite fields (1.1-3) in a unified manner we set

$$\text{(A.1)} \qquad O_{n+k}(x,z) = :\phi^*(x)\, M_k(z)\, \mathcal{D}_n(z\overleftarrow{\nabla}\, ,\, z\overrightarrow{\nabla})\phi(x): \quad (\mathcal{D}_n = \mathcal{D}_n^{\nu})$$

where $\phi(x)$ is a spin-tensor field carrying spin $s = \frac{1}{2}k$ (for 0-mass fields we identify the spin with the absolute value of the helicity), M_k is a matrix-valued homogeneous polynomial of z of degree k and $D_n(\alpha,\beta)$ is a homogeneous polynomial of α and β of degree n. We assume that the field ϕ is conformally covariant with infinitesimal special conformal transformation law

$$\text{(A.2)} \qquad \frac{1}{2i}\,[\phi(x),K_\mu] = (x_\mu(d + x\nabla) - \frac{x^2}{2}\,\nabla_\mu + ix^\nu\, S_{\mu\nu})\,\phi(x);$$

here d is the scale dimension of ϕ, $S_{\mu\nu}$ are the finite dimensional Lorentz generators:

$$\text{(A.3a)} \qquad S_{\mu\nu} = \frac{1}{4i}\,[\gamma_\mu,\gamma_\nu] = \frac{1}{2}\begin{pmatrix}\sigma_{\mu\nu} & 0\\ 0 & \sigma^*_{\mu\nu}\end{pmatrix} \qquad \begin{aligned}\sigma_{\mu\nu} &= \frac{i}{2}\,(\sigma_\mu\tilde{\sigma}_\nu - \sigma_\nu\tilde{\sigma}_\mu)\\ \sigma_0 &= \tilde{\sigma}_0 = \begin{pmatrix}1&0\\0&1\end{pmatrix},\ \sigma_i = -\tilde{\sigma}_i\end{aligned}$$

(the second equation is valid in the γ_5-diagnonal realization in which $\gamma_\mu = -i\begin{pmatrix}0 & \sigma_\mu\\ \tilde{\sigma}_\mu & 0\end{pmatrix}$) in the case of a Dirac field,

$$\text{(A.3b)} \qquad -i(S_{\mu\nu})_\lambda^\kappa = \eta_{\mu\lambda}\delta_\nu^\kappa - \eta_{\mu\lambda}\delta_\mu^\kappa \qquad \text{for a vector field}$$

$$\text{(A.3c)} \qquad S_{\mu\nu} = -i(z_\mu\partial_\nu - z_\nu\partial_\mu) \ \text{for the field}\ O_\ell(x,z).$$

We further require the covariance property

(A.3d) $\quad S^{*}_{\mu\nu} M_k(z) = M_k(z) S_{\mu\nu} + i(z_\mu\partial_\nu - z_\nu\partial_\mu)M_k(z)$.

The arguments $\alpha = z\overleftarrow{\nabla}$ and $\beta = z\overrightarrow{\nabla}$ of $\mathcal{D}_n$ may involve either the ordinary derivative ∇_μ (acting to the left and to the right) - in the case of a scalar constituent field φ, or the covariant derivative $\mathcal{D}_\mu$(1.5). For a given M_k we shall derive the expression for $\mathcal{D}_n$ by the requirement that O_ℓ is a(local) conformal covariant field, such that the stabilizer of the point $x = 0$ in the conformal group is represented irreducibly. This implies triviality of infinitesimal special conformal transformations at $x = 0$:

(A.4) $\quad [O_\ell(x,z),\mathcal{K}_\nu]|_{x=0} = 0.$

Applying (A.4) to (A.1,2) and observing the properties

(A.5a) $\quad x_\nu(d+x\nabla)\phi^{*}M_k(z)\mathcal{D}_n(\alpha,\beta)|_{x=0} = z_\nu(d + \alpha\frac{\partial}{\partial\alpha})\phi^{*}M_k\frac{\partial\mathcal{D}_n}{\partial\alpha}$,

(A.5b) $\quad \mathcal{D}_n(\alpha,\beta)x_\nu(d+x\nabla)\phi(x)|_{x=0} = z_\nu(d+\beta\frac{\partial}{\partial\beta})\frac{\partial\mathcal{D}_n}{\partial\beta}\phi$,

(A.5c) $\quad x^2\mathcal{D}_n(\alpha,\beta)|_{x=0}, = \mathcal{D}_n(\alpha,\beta)x^2|_{x=0} = 0,$

we find

(A.6) $$\frac{1}{2i}[O_{n+k}(x,z), K_\nu]_{x=0} = z_\nu:\phi^{*}M_k(z)\{(d+s)(\frac{\partial}{\partial\alpha} + \frac{\partial}{\partial\beta}) + \alpha\frac{\partial^2}{\partial\alpha^2} + \beta\frac{\partial^2}{\partial\beta^2}\}\mathcal{D}_n(\alpha,\beta)\phi: = 0$$

$(\alpha = z\overleftarrow{\nabla},\quad \beta = z\overrightarrow{\nabla},\quad s = \frac{k}{2})$.

The term involving s comes from the relation

(A.7) $$i(x^\mu S^*_{\mu\nu} M_k(z)\mathcal{D}_n(\alpha,\beta)-\mathcal{D}_n(\alpha,\beta)M_k(z)x^\mu S_{\mu\nu}) =$$

$$= z_\nu \frac{1}{2} z\partial M_k(z) \left(\frac{\partial}{\partial\alpha} + \frac{\partial}{\partial\beta}\right)\mathcal{D}_n(\alpha,\beta)= \frac{k}{2}z_\nu M_k(z) \left(\frac{\partial}{\partial\alpha} + \frac{\partial}{\partial\beta}\right)\mathcal{D}_n$$

which follows from (A.3d) and from the homogeneity of M_k.
Eq. (A.6) is satisfied for arbitrary ϕ iff

(A.8a) $$\left[\nu\left(\frac{\partial}{\partial\alpha} + \frac{\partial}{\partial\beta}\right) + \alpha\frac{\partial^2}{\partial\alpha^2} + \beta \frac{\partial^2}{\partial\alpha^2}\right]\mathcal{D}^\nu_n(\alpha,\beta) = 0$$

(A.8b) $$= \left[(\nu + n-1)\left(\frac{\partial}{\partial\alpha} + \frac{\partial}{\partial\beta}\right) - (\alpha + \beta) \frac{\partial^2}{\partial\alpha\partial\beta}\right] \mathcal{D}^\nu_n(\alpha,\beta) \quad (\nu= d+s)\cdot$$

(The second of these equations makes use of the implication

(A.9) $$\left(\alpha\frac{\partial}{\partial\alpha} + \beta\frac{\partial}{\partial\beta} - n+1\right) \left(\frac{\partial}{\partial\alpha} + \frac{\partial}{\partial\beta}\right) \mathcal{D}^\nu_n(\alpha,\beta) = 0$$

of the homogeneity condition for $\mathcal{D}_n$.) Inserting (1.3b) in (A.8) we obtain

(A.10) $$\left[(1-\xi^2) \frac{d^2}{d\xi^2} - 2\nu\xi \frac{d}{d\xi} + n(n+2\nu-1)\right] P_n(\xi) = 0 \quad \left(\xi = \frac{\beta-\alpha}{\alpha+\beta}\right).$$

The general polynomial solution of (A.10) is proportional to a Gegenbauer polynomial:

(A.11) $$P_n(\xi) = N^\nu_n C^{\nu-\frac{1}{2}}_n (\xi).$$

In the special case of Dirac field $k = 2s = 1$,

(A.12a) $\quad M_1(z) = \beta\Gamma z$

($\beta = \beta^*$ is the Dirac conjugation matrix: $\tilde{\psi} = \psi^*\beta$, $S^*_{\mu\nu}\beta = \beta S_{\mu\nu}$); for a gluon field

(A.12b) $\quad (M_2(z)G_a)^{\mu\nu} = \frac{1}{2}(z^\mu G_a^{\mu\rho} - z^\nu G_a^{\mu\rho})\, z_\rho .$

In both cases the free equations of motion are given by

(A. 13) $\quad \nabla\delta M_k(z)\phi(x) \equiv (k\nabla\partial - \frac{z\nabla}{2}\,\partial^2)M_k(z)\phi(x) = 0$

where δ is the interior derivative on the cone $z^2 = 0$ given by (1.6b). It can be verified that the free field equation (A.13) implies the conservation law (1.6) for the composite field (A.1).

We note that if ϕ is a hermitian field (like in the gluon case) then the odd degree O_ℓ vanish since

(A.14) $\quad C_n^{\nu-\frac{1}{2}}(-\xi) = (-1)^n\, C_n^{\nu-\frac{1}{2}}(\xi)$ implies $\mathcal{D}_n(\alpha,\beta) = (-1)^n\mathcal{D}_n(\beta,\alpha)$.

Appendix B

Evaluation of triangular diagrams for composite operators

We shall derive the representation (2.5) for the contribution $I_{(A)}$ of the triangular graph of Fig. 1A in the case of scalar fields and will then indicate the necessary modification when Dirac spinor fields are involved.

Using the standard representation we can write the (Euclidean, p-space) expression for $I_{(A)}$ as

(B.1) $$I_{(A)}(p_1,p_2;z) = -g^2 \int_0^\infty dx_1 \int_0^\infty dx_2 \int_0^\infty dx_3 \int d_{2h}k \, \mathcal{D}_\ell^{h-1}(z(p_1-k), z(k-p_2)) \times e^{-\alpha_1(p_2-k)^2 - \alpha_2(p_1-k)^2 - \alpha_3 k^2} =$$

$$= \frac{g^2}{(4\pi)^2} \iiint_0^\infty \frac{\prod_1^3 d\alpha_i}{(\alpha_1+\alpha_2\ \alpha_3)^h} \mathcal{D}_\ell^{h-1}\left(-\frac{z}{2\alpha_2}\frac{\partial}{\partial p_1}, \frac{z}{2\alpha_1}\frac{\partial}{\partial p_2}\right) \exp\left\{-\frac{\alpha_2\alpha_3 p_1^3 + \alpha_2\alpha_3 p_2^2 + \alpha_1\alpha_2(p_1-p_2)^2}{\alpha_1+\alpha_2+\alpha_3}\right\}_2$$

$$\left(d_{2h}k = \frac{d^{2h}k}{(2\pi)^{2h}}\right).$$

(In the replacing the arguments of $\mathcal{D}_\ell$ with the corresponding derivatives we have noted that $z^2 = 0$.) Next we evaluate the derivatives and change the integration variables setting

(B.2a) $$\alpha_1 = \lambda\alpha\beta, \quad \alpha_2 = \lambda(1-\alpha)\beta, \quad \alpha_3 = \lambda(1-\beta) \quad (\lambda = \alpha_1+\alpha_2+\alpha_3),$$

so that

(B.2b) $d\alpha_1 d\alpha_2 d\alpha_3 = \lambda^2 \beta d\lambda d\beta d\alpha \quad (0 \le \alpha \le 1,\ \ 0 \le \beta \le 1,\ 0 \le \lambda < \infty)$.

The result is

$$I_{(A)} = -\frac{g^2}{(4\pi)^h}\int_0^1 d\alpha \int_0^1 \beta d\beta \int_0^\infty \lambda^{2-h} d\lambda\, \mathcal{D}_\ell^{h-1}\,(\,z\{(1-\beta)p_1+\alpha\beta(p_1-p_2)\}\,,$$

$$-z\{(1-\beta)p_2-(1-\alpha)\beta(p_1-p_2)\}\times \times\, e^{-\lambda\beta\{(1-\beta)[(1-\alpha)p_1^2 + \alpha p_2^2\,] + \alpha(1-\alpha)\beta(p_1-p_2)^2\}}$$

(B.3)

$$= -\frac{g^2}{(4\pi)^h}\,\Gamma(3-h)\int_0^1 d\alpha \int_0^1 \beta^{h-2} d\beta\, \times$$

$$\times\,\frac{\mathcal{D}_\ell^{h-1}(z\{1-\beta)p_1+\alpha\beta(p_1-p_2)\},\ -z\{(1-\beta)p_2+(1-\alpha)\beta(p_2-p_1)\})}{\{(1-\beta)[(1-\alpha)p_1^2+\alpha p_2^2] + \alpha(1-\alpha)\beta(p_1-p_2)^2\}^{3-h}}\,.$$

Setting $3-h = \varepsilon \to 0$ and evaluating the coefficient to $\frac{1}{\varepsilon}$ we obtain (2.5).

For a Dirac field interacting with a gauge field the integrand in (B.1) should be multiplied in the Feynman gauge by $-\gamma^\rho(\not{p}_2-\not{k})i\gamma z(\not{k}-\not{p}_1)\gamma_\rho$, or equivalently by the operator

(B.4) $$\gamma^\rho \frac{1}{2\alpha_1}\,\gamma\frac{\partial}{\partial p_2}\, i\gamma z\, \frac{1}{2\alpha_2}\,\gamma\frac{\partial}{\partial p_1}\,\gamma_\rho\,.$$

The leading $(\frac{1}{\varepsilon})$ term in the counterpart of $I_{(A)}$ is

$$I_{(A)}^{(G)}\,(p_1,p_2;z) \simeq -C_2(G)\frac{g^2}{(4\pi)^h}\iiint_0^\infty \frac{\Pi d\alpha}{(\alpha_1+\alpha_2+\alpha_3)^{h+1}}\,\mathcal{D}_\ell^h\,(-\frac{z}{2\alpha_2}\,\frac{\partial}{\partial p_1},\frac{z}{2\alpha_1}\,\frac{\partial}{\partial p_2})\times$$

(B.5)

$$\times\,\frac{i}{2}\,\gamma^\rho\gamma^\lambda\gamma_z\gamma_\lambda\gamma_\rho\ \ e^{-\frac{\alpha_2\alpha_3 p_1^2+\alpha_1\alpha_3 p_2^2+\alpha_1\alpha_2(p_1-p_2)^2}{\alpha_1+\alpha_2+\alpha_3}}$$

$$\simeq -\frac{C_2(G)}{(4\pi)^2}\,\Gamma(2-h)g^2 i\gamma z\int_0^1 d\alpha\, 2\int_0^1 \beta^{h-1}d\beta\;\frac{\mathcal{D}^h_\ell(z\{(1-\beta)p_1+\alpha\beta(p_1-p_2)\},-z\{(1-\beta)p_2+(1-\alpha)\beta(p_2-p)\})}{\{(1-\beta)[(1-\alpha)p_1^2+\alpha p_2^2\,]+\alpha(1-\alpha)\beta(p_1-p_2)^2\}^{2-h}}$$

where $C_2(G)$ is the second order Casimir operator of the gauge group

(B.6) $\quad C_2(SU(N)) = \frac{N^2-1}{2N}$ for $N = 2,3$, $C_2(U(1) = 1$

and $\varepsilon = 2-h(\to 0)$.

Appendix C

Preservation of gauge and conformal invariance of composite operators under renormalization

We shall demonstrate that renormalization of second rank tensor composite operators respects gauge invariance and reproduces the conformal structures. To fix the ideas, we shall only consider the 2-photon contribution to δO, where O stands for either O^F or O^ψ:

$$\text{(C.1)} \qquad \delta O(x,z) = \iint dy_1 dy_2\, \Gamma^{\mu_1\mu_2}(x;y_1,y_2;z):A_{\mu_1}(y_1)A_{\mu_2}(y_2): \;,$$

$\Gamma^{\mu_1\mu_2}$ being the 1-particle irreductible vertex function $\langle O(x,z)j^{\mu_1}(y_1)j^{\mu_2}(y_2)\rangle_E^{1p_i}$.
The Fourrier transform $\tilde{\Gamma}^{\mu_1\mu_2}$, defined by

$$(2\pi)^4\delta(p+q_1+q_2)\tilde{\Gamma}^{\mu_1\mu_2}(q_1,q_2;z) = \iiint dx dy_1 dy_2 \Gamma^{\mu_1\mu_2}(x;y_1,y_2;z)e^{-ipx-iq_1y_1-iq_2y_2}$$

satisfies the transversality condition (or the Ward identity)

$$\text{(C.2)} \; q_{1\mu_1}\tilde{\Gamma}^{\mu_1\mu_2}(q_1,q_2;z) = 0 = q_{2\mu_2}\tilde{\Gamma}^{\mu_1\mu_2}(q_1,q_2;z).$$

This condition will be shown to imply the proportionality of the singular parts of δO to O^F.

The most general Lorentz invariant symmetric vertex function can be written as

$$\tilde{\Gamma}^{\mu_1\mu_2}(q_1,q_2;z) = q_1^{\mu_1}q_2^{\mu_2} A_1(q_1,q_2;z) + q_2^{\mu_1}q_1^{\mu_2} A_2(q_1,q_2;z) +$$

$$+ q_1^{\mu_1}z^{\mu_2}B_1(q_1,q_2;z) + z^{\mu_1}q_2^{\mu_2}B_1(q_2,q_1;z) +$$

$$+ q_1^{\mu_2}z^{\mu_1} B_2(q_1,q_2;z) + q_2^{\mu_1}z^{\mu_2}B_2(q_2,q_1;z) + \tag{C.3}$$

$$+ z^{\mu_1}z^{\mu_2}C(q_1,q_2) + \eta^{\mu_1\mu_2}D(q_1,q_2;z) + q_1^{\mu_1}q_1^{\mu_2}E_1(q_1,q_2;z) +$$

$$+ q_2^{\mu_1},q_1;z);$$

here all coefficients are numerical functions of the inner products, and A_i,C and D are symmetric in q_1,q_2:

$$A_i(q_1,q_2;z) = A_i(q_2,q_1;z), \quad C(q_1,q_2) = C(q_2,q_1), D(q_2,q_1;z).$$

The transversality relation (C.2) yields (skipping the argument z in A_i,B , D and E

$$q_1^2 A_1^{(q_1,q_2)} + zq_1B_1(q_2,q_1) + q_1q_2 E(q_2,q_1) = 0 \tag{C.4}$$

$$q_1q_2A_2(q_1,q_2) + zq_1B_2(q_1,q_2) + D(q_1,q_2) + q_1^2 E(q_1,q_2) = 0$$

$$q_1^2 B_1(q_1,q_2) + q_1q_2B_2(q_2,q_1) + q_1zC(q_1,q_2) = 0.$$

The $\frac{1}{\varepsilon}$ part of $\Gamma^{\mu_1\mu_2}$ is a second degree polynomial in the momenta. This determines the $\frac{1}{\varepsilon}$ terms in the coefficients in (C.3) up to a few constants. Indeed, taking into account the assumption that O(x) is a second rank tensor so that each term in the right hand side of (C.3) is homogeneous of degree 2 in z (and that $z^2 = 0$) we find:

$$A_i^{(\varepsilon)} = 0 = E, \; B_i^{(\varepsilon)} = \frac{1}{\varepsilon} B_{ij}(zq_j)$$

(C.5) $$C^{(\varepsilon)} = \frac{1}{\varepsilon}(C_1 \frac{q_1^2 + q_2^2}{2} + C_2 q_1 q_2),$$

$$D^{(\varepsilon)} = \frac{1}{2\varepsilon}\{D_1[(zq_1)^2 + (zq_2)^2] + 2D_2(zq_1)(zq_2)\}.$$

Inserting (C.5) in (C.4) we obtain

(C.6) $$B_1^{(\varepsilon)} = 0 = B_{21} = D_1 = C_1, \quad D_2 = C_2 = -B_{22} \equiv C,$$

so that a single structure survives in the divergent part of $\tilde{\Gamma}^{\mu_1\mu_2}$:

(C.7) $$\tilde{\Gamma}^{\mu_1\mu_2}(q_1,q_2;z) = \frac{C}{\varepsilon}(z^{\mu_1}z^{\mu_2}q_1q_2 + zq_1zq_2\eta^{\mu_1\mu_2} - q_2zq_1{}^{\mu_2}z^{\mu_1} - q_1zq_2{}^{\mu_1}z^{\mu_2}) .$$

Consequently (noting that $O^F = z^\mu F_{\mu\rho}F^{\rho\nu}z_\nu = 2(z\nabla A^\rho)\nabla_\rho zA - (\nabla_\rho zA)\nabla^\rho zA - z\nabla A^\rho z\nabla A_\rho$) we find

(C.8) $$\delta O(x,z) = \frac{C}{\varepsilon} O^F(x,z) + \text{finite terms (for } \varepsilon \to 0).$$

In order to evaluate the element C of the anomalous dimension matrix it suffices to compute the coefficient to one of the structures (say to $z^{\mu_1}z^{\mu_2}$).

For a recent study of renormalization of composite operators in gauge theories see [21].

REFERENCES

(1) S. FERRARA, R.GATTO, A.GRILLO, G. PARISI, General consequences of conformal algebra, in: Scale and Conformal Symmetry in Hadron Physics, ed. by R. Gatto (Wiley, New York 1973) pp. 59-108); G. MACK, Conformal invariant quantum field theory, ibid. pp. 109-130.

(2) I.T. TODOROV, M.C. MINTCHEV, V.B. PETKOVA, Conformal Invariance in Quantum Field Theory (Scuola Normale Superiore, Pisa 1978).

(3) E.S. FRADKIN, M.Ya. PALCHIK, Recent developments in conformal invariant QFT, Phys.Reports 44C , 249-349 (1978).

(4) A.V. EFREMOV, A.V. RADYUSHKIN, Asympotitic behavior of the pion form factor in quantum chromodynamics, Teor.Mat.Fiz.42, 147-166(1980), (transl.: Theor. Math. Phys. 42, 97-110(1980)); see also JINR preprint E2-11983, Dubna (1978).

(5) V.K. DOBREV, V.B. PETKOVA, S.G. PETROVA, I.T. TODOROV, Dynamical derivation of vacuum operator product expansion in Euclidean conformal quantum field theory, Phys.Rev. D13, 887-912(1976).

(6) S.J. BRODSKY, Y. FRISHMAN, G.P. LEPAGE, C. SACHRAJDA, Hadronic wave functions at short distances and the operator product expansion, Phys.Letters 31B, 239-244 (1980).

(7) N.S. CRAIGIE, J. STERN, What can we learn from sum rules involving vertex functions in QCD, Nucl. Phys. B216, 209-243 (1983).

(8) N.S. CRAIGIE, V.K. DOBREV, I.T. TODOROV, Conformal techniques for OPE in asymptotically free quantum field theory, ICTP, Trieste, preprint IC/82/63; Conformally covariant composite operators in quantum chromodynamics, ICTP, Trieste, preprint IC/83/35.

(9) A.A. MIGDAL, Multicolor QCD as a dual resonance theory. Annals of Physics 109, 365-392, (1977); Yu.M.MAREENKO, On conformal operators in QCD, Yad.Fiz.33, 842-847 (1981) (transl.: Sov.J.Nucl. Phys. 33, 440-442(1981))

(10) Th.OHRNDORF, Constraints from conformal invariance on the mixing of operators of lower twist, Nucl.Phys. B198, 26-44 (1982).

(11) V.B. PETKOVA, G.SOTKOV, I.T.TODOROV, Local field representations of the conformal group and their physical interpretation (lecture presented at the International School on Differential Geometric Methods in Mathematical Physics, Varna, Bulgaria 1982); see also V.B. PETKOVA I.T. TODOROV, ISAS preprint 14/83/E.P., Trieste.

(12) A.I. OKSAK, I.T.TODOROV, On the covariant structure of the two-point function, Commun.Math.Phys. 14, 271-304 (1969); I.T. TODOROV, R.P. ZAIKOV Spectral representation of the co-variant two-point function and infinite composite fields with arbitrary mass spectrum, J. Math. Phys. 10, 2014-2010 (1969)

(13) V. BARGMANN, I.T. TODOROV, Spaces of analytic functions on a complex cone as carriers for the symmetric tensor representations of SO(n), J.Math.Phys. 18, 1141-1148 (1977).

(14) V.K. DOBREV, G.MACK, V.B.PETKOVA, S.G.PETROVA, I.T. TODOROV, Harmonic Analysis on the n-Dimensional Lorentz Group and its Application to Conformal Quantum Field Theory, Lecture Notes in Physics 63 (Springer, Berlin 1977).

(15) S.A. ANIKIN and O.I. ZAVIALOV, Short distance and light-cone expansions for products of currents, Ann. Phys.(N.Y.)116, 135-166(1978).

(16) B. BINEGAR, C. FRONSDAL, W. HEIDENREICH, Conformal QED, J.Math.Phys. 24, 2828-2846 (1983).

(17) R.P. ZAIKOV, On the conformal invariance in gauge theories: quantum electrodynamics, JINR preprint E2-83-28, Dubna (1983)

(18) P. FURLAN, V.B. PETKOVA, G.M. SOTKOV, I.T. TODOROV, Conformal quantum electrodynamics with a 5-potential, ISAS preprint 52/83/E.P., Trieste.

(19) M. BAKER, K. JOHNSON, Application of conformal symmetry in quantum electrodynamics, Physica 96A, 120-130 (1979).

(20) N. CHRIST, B. HASSLACHER , A. MUELLER, Light-cone Behavior of Perturbation Theory. Phys.Rev. D6, 3543-3562, (1972), D.J.GROSS, Applications of the renormalization group to high-energy physics, in: Methods in Field Theory, Eds. R. Balian and J. Zinn-Justin, Les Houches, Session XXVIII, 1975 (North Holland, Amsterdam 1976) pp. 140-250.

(21) A. ANDRAŠI, J.C. TAYLOR, Renormalization of composite operators in gauge theories, Nucl. Phys. B227, 494-502 (1983).